Manufactured
Uncertainty

Manufactured Uncertainty

Implications for Climate Change Skepticism

Lorraine Code

Published by State University of New York Press, Albany

© 2020 State University of New York

All rights reserved

Printed in the United States of America

No part of this book may be used or reproduced in any manner whatsoever without written permission. No part of this book may be stored in a retrieval system or transmitted in any form or by any means including electronic, electrostatic, magnetic tape, mechanical, photocopying, recording, or otherwise without the prior permission in writing of the publisher.

For information, contact State University of New York Press, Albany, NY
www.sunypress.edu

Library of Congress Cataloging-in-Publication Data

Name: Code, Lorraine, author.
Title: Manufactured uncertainty : implications for climate change skepticism / Lorraine Code.
Description: Albany : State University of New York Press, [2020] | Includes bibliographical references and index.
Identifiers: LCCN 2020018292 | ISBN 9781438480534 (hardcover : alk. paper) | ISBN 9781438480541 (pbk. : alk. paper) | ISBN 9781438480558 (ebook)
Subjects: LCSH: Climatic changes—Philosophy. | Environmental responsibility. | Knowledge, Sociology of.
Classification: LCC QC903 .C59 2020 | DDC 363.738/7401—dc23
LC record available at https://lccn.loc.gov/2020018292

10 9 8 7 6 5 4 3 2 1

For the next generation:

Michaela Code

Graeme Playfair

Sophia Code

Iain Playfair

Aubrey McCance

. . . in their quests for the pleasures and promises of certainty in uncertain times

Contents

Foreword	ix
Acknowledgments	xi
Introduction	1
Chapter 1 Epistemic Responsibility, Now	27
Chapter 2 Doubt and Denial: Epistemic Responsibility Meets Climate Change Skepticism	63
Chapter 3 Care, Concern, and Advocacy: Is There a Place for Epistemic Responsibility?	99
Chapter 4 Particularity, Epistemic Responsibility, and the Ecological Imaginary	141
Chapter 5 How to Think Globally, Revisited: Or, A Plea for Ignorance	177
Bibliography	219
Index	229

Foreword

The essays collected here address a cluster of puzzles, possibilities, and conundrums that emerge in and from current Anglo-American epistemological practices.

They derive from thinking about responsible epistemic conduct in the twenty-first century Western/Northern urban world, where questions about objectivity and abstraction are claiming less attention than they once did and where more concrete epistemic practices—the local specificities of knowing in and about the world—claim a new salience. Matters of objectivity, separated from quotidian practices of knowing and understanding well, are less central to the issues addressed in these essays than are questions about how to reconceive the ancient idea of "virtuous knowledge" for lives diversely situated within the contested social-political orders of the Anglo-American urban world as we now find it. In such inquiry, the irreducible particularities of human lives and situations demand empathetic attention as the traditionally abstract epistemic agent recedes from view, and diverse, hitherto marginalized ways of knowing urgently claim a key role in contemporary debate.

—Lorraine Code

Acknowledgments

I owe thanks and gratitude to so many who have participated in bringing this book to completion.

I am indebted to the Social Sciences and Humanities Research Council of Canada (SSHRCC) for the research grant that supported my work throughout the book.

Jamie Robertson's intelligent and capable participation as a research assistant throughout its writing and production contributed immeasurably to its completion. Without her assistance it would not have emerged in its present form.

Paula Butler prepared the index for this volume.

My son David Code has provided invaluable editorial assistance and thoughtful advice from beginning to end, and Jacqueline Code's moral support has seen me through moments of doubt.

Murray Code, as always, has made its final emergence possible.

Introduction

The purpose of this book is to engage with some of the most urgent epistemological and ontological effects of public skepticism and doubt in relation to climate change, in the twenty-first-century Western-Northern world and beyond, viewing these effects through a lens focused on entrenched but often misguided assumptions about how such matters can responsibly be known, lived, engaged. Harms across a broad social, political, demographic, and geographic spectrum consequent upon industrial pollution, tobacco smoking, global warming, ecological devastation, and a vast range of interconnected private and public practices encouraged and condoned by such doubts, are much in the public eye in the twenty-first century. They are impressively analyzed and debated in Naomi Oreskes's and Eric Conway's 2010 book *Merchants of Doubt*: a text that figures centrally among the catalysts for this engagement with such practices.[1] It is the source of the title: "Manufactured Uncertainty," which is intended as a challenge to socially embedded assumptions about the taken for granted "certainty"—the rarely contested "reliability"—of publicly announced and analysed scientific findings related to "environmental"/ecological matters in many twenty-first-century Western-Northern societies and academic institutions.

The term "environmental" is itself contestable in that it centers "us"—we human beings—who are the very people who invoke it: hence the reflections to follow favor a language of ecology/ "ecological," wherever

---

[1]. Naomi Oreskes and Erik M. Conway, *Merchants of Doubt: How a Handful of Scientists Obscured the Truth on Issues from Tobacco Smoke to Global Warming* (London: Bloomsbury, 2010).

it is feasible/plausible.[2] Most significantly, I intend to show, such ongoing harms attest to deep-seated social-political-ontological assumptions about who we are—we who pose such questions—and about how we understand and enact our place(s) in the world. Their significance for feminist-, race-, place-, class-specific and other interconnected areas of twenty-first-century theory and practice remains underexamined: it is here that many of these interventions will focus. When skeptics appeal to a margin of uncertainty in climate change science, with the aim of generating and fueling incredulity and justifying resistance to regulating social and individual behaviors, they commonly defend a conception of freedom—of Liberty—assumed uncontestably to be theirs, to pertain universally, and to be unjustly threatened by contentions that they cannot/should neither consume nor pollute as they will. Hence this text engages with some of the damaging effects of lived assumptions about freedom and entitlement in the twenty-first-century Western world: assumptions which vary across human and situational-circumstantial diversity, yet whose cumulative effects are reciprocally reinforcing. These engagements invoke ethical-political questions about epistemic responsibility (a responsibility to know carefully and well): they prompt fundamental yet diverse ontological questions about who "we" are and how "we" can live responsibly, together and separately, in the human, natural, and social-political world. A principal goal of this project, then, is to generate a rethinking of (often tacit) socially and philosophically entrenched convictions about universal human sameness, about "our" entitlement to consume and pollute as "we" will, and about responsibilities—epistemic responsibilities—to know carefully and well. A guiding hope is that it will prompt educators and activists to engage in deliberations/disputations that could animate radical shifts in the sedimented policies and practices that govern quotidian and public-political action in the affluent white Western world, especially where (contestable) assumptions about human "sameness" prevail, too often with coercive consequences.

In short, these thoughts generate probing questions about *who "we" think we are*: we who write, and think, and speak as diversely educated and multiply privileged members of affluent twenty-first-century societies:

---

2. For an excellent analysis of the significance of speaking of "ecological" as contrasted with "environmental" thinking, see M. Hazlett, "'Woman vs. Man vs. Bugs': Gender and Popular Ecology in Early Reactions to Silent Spring," *Environmental History* 9 no. 4 (Oct. 2004), pp. 701–29.

we who, across a wide range of significant senses, have made the world for which we are now accountable. "We" need urgently to find/discover/craft ways of restoring and preserving that world for people who are not so fortunate: for the ignored and oppressed; for future people, species, and less affluent, neither secure nor privileged populations who must live with the effects—the consequences—be they positive or negative, of "our" ways of living. Here, social justice is an overarching issue. Thus, the framing of this inquiry is ethical-epistemological-ontological-political. It is articulated from a position shaped by feminist and postcolonial engagements in recognizing that knowing responsibly and well, singly or collectively, is *sine qua non* for responsible social action *now*—where "responsible" incorporates responsiveness and recognition in engaging with people, places, practices, theories, things, and situations "on their own terms," so far as this is possible. A central contrast is with practices of superimposing categories and explanations that subsume differences under preformed classifications, with inadequate attention to how an appropriate "fit" could or should be achieved. In its commitments to thinking communally, socially, collaboratively, cooperatively—*even contrarily*—across human, other than human, and situational differences, the position I will articulate eschews the radical individualism of Western philosophy and social theory to focus on knowing/knowledge making as a cluster of socially enacted practices.

Rethinking and reenacting *who we are* in ways sufficiently powerful to dislodge these sedimented convictions is the hardest yet the most urgent task for this inquiry, as it is, if perhaps tacitly, throughout the world as we find it. To practice such a philosophy and develop such pedagogical practices requires—*must* require—engaged questioning that is ontologically-epistemologically radical, challenging—upheld as those assumptions are by the instituted social-political-epistemic imaginary in which we (diversely privileged) inhabitants of the affluent West/North live and think and have our being, however obliquely or contrarily. Engaging seriously in these thoughts and the actions they inspire calls for crafting, and living, critically renewed conceptions of human subjectivity across a range of personal, social, political, collective levels, while reconstructing sedimented conceptions of ethical-political engagement in/with the other-than-human world, both animate and inanimate. Requirements such as these are distantly analogous to late-twentieth-century Western-Northern consciousness-raising practices in that they call for critical genealogical analyses of who "we" are and how "we" are accustomed to live: analyses conducted in ways sufficiently discerning to unsettle fixed expectations that inform many of "our" everyday thoughts and actions. They must be sufficiently radical to engage historically and cross-culturally with the "absolute presuppositions," thus named by R. G. Collingwood, critiquing the neoempiricist assumptions of early analytic philosophy, in their "trickle-down" effects in/for twenty-first-century lives.[3] This is the hardest requirement: it is easier to engage in revisionary ways of doing, thinking, knowing. But to practice a philosophy that requires unsettling and reconceiving basic assumptions about *who we are* is ontologically epistemologically radical, upheld as these assumptions are by an instituted social-political-epistemic imaginary (following Cornelius Castoriadis) in which "we" inhabitants of the West live and think and enact our *Being*, however obliquely or contrarily. In this respect, the project has a particular bearing on educational practice, for "we" who are educators carry a special responsibility to know such issues responsibly and well, to allow that knowledge to inform our pedagogical practices, to open inquiry for those who live alongside us, depend upon us, and come after us.

I am suggesting that projects engaged in exploring interconnections between "human rights and the environment" are commonly impeded by locating their point of departure so high up on a quasivertical explanatory structure that they presuppose too much and truncate inquiry in so doing. In failing to start from the ground up, from engaging critically with the enactments of human subjectivity they unquestioningly presuppose, such inquiries too often produce insufficiently radical analyses and understandings. Here, then, a prerequisite to living well with the complexities of climate change skepticism and its multiple, multiply uneven implications for social justice is a critical-creative engagement with deeply embedded yet often uncontested assumptions about human subjectivity and agency as these silently inform the social imaginary of the white affluent Northern-Western world, and as they tacitly shape ecological/environmental thinking in its social-political effects. Such assumptions run counter to the substance and manner of the ecological thinking and engagements I advocate.

Skepticism in the iterations to be addressed here covers a range of practices and attitudes that stretch from outright denial that there could

---

3. See R. G. Collingwood, *An Essay on Metaphysics* (Oxford: Oxford University Press, 1939), who advocates a transformation of metaphysics from a study of being or ontology to a study of the absolute presuppositions or heuristic principles that govern fixed ways of being and knowing.

be any knowable cause for concern about the state of the world—"-nothing to worry about"—through to insistent contentions that productive ecological debate needs to take as its point of departure the fundamental, challenging question "*Who* do we think we are?" A tacit assumption prevails in common parlance and in philosophical discourse alike that "we" know quite well who we are and, by extension, know the scope and limits of the rights, needs, and entitlements consequent upon claiming, and enacting, these "identities." Often, without further inquiry, such judgments presuppose a settled, complacent, quasi-essentialist understanding of what it is to be (generically) human, in itself and in its actions, relations, and situations: an understanding that is neither ethically nor epistemologically innocent nor immune to challenge. It is vital to revisit and interrogate it because such settled, uncontested assumptions rarely take widespread, warranted (= healthy) skepticism into account. Even though presupposition-free analysis is surely impossible, a quasi-genealogical analysis of the discernible constitutive assumptions that animate a line of inquiry and a settled way of life can be invaluable in establishing a point of entry into investigations that risk presupposing too much from the outset. These are among the questions, I suggest, that should animate philosophical-ecological deliberations at multiple levels—ontological, epistemological, ethical, political—in working toward developing conceptions of viable ecological theory and practice and of ecological subjectivity. Here, feminist inquiry in its diverse iterations claims a significant place among issues that call for disputation and resistance. The skepticism—frequently virulent—that has fueled blatantly (if covertly) oppressive ways of life, sustaining exclusionary patterns and practices of social-political life and policy, has generated decades—centuries—of purposefully ignorant ageist, sexist, racist oppression throughout the Western/Northern world: it has worked cruelly to exclude unimaginable numbers of people from the very possibility of "being human." In many such iterations, skepticism can be harmful, individually and collectively. Yet, skepticism is multifaceted, complex, and also often—if paradoxically—healthy. In its quotidian iterations, it commonly manifests, quite simply, in persistent and/or episodic doubt and denial about trivial or complex issues ranging from reluctance/refusal to believe a range of everyday empirical knowledge claims, through to socially-politically complex, frequently fraught affirmations, challenges, and policies.

Pertinent for the discussions throughout this book is *climate change skepticism* in its multiple modalities. The issue is visible and palpable in a diverse range of doubt and denial across practices and populations, together with a range and variety of uncontested practices—and of collective "self-denial"—that permit these practices to thrive. A plausible (if self-centered) back story focuses on the losses, be they social, material, or selfish—that certain denials enable. In the twenty-first century, the "haves" are deeply, even ferociously, reluctant to have less. But there are more complex forms and patterns of resistance/reluctance, more plausibly uncertain beliefs and practices. Hence, to the contrary, some skeptics (at times with justification) are convinced that even well-informed resisters underestimate the insistent threats of climate change and hence are frustrated when/if they encounter stubborn incredulity. There are those who resist the unanimity that may fuel critical resistance, or who worry about a consensus emerging from tendencies to aggregate disparate issues, with the consequence that vital-to-address irregularities, anomalies, disappear into the (admittedly urgent) generalities. Clearly, none of these epistemic practices can promise certainty, but their enactments need to be held open to disputation: to keep inquiry alive and alert. Issues of certainty and uncertainty are germane to such practices as they are enacted, or acted upon, in the wider world. While professed uncertainty may be an energizing and/or a frustrating, practice-shaping concept, in the positivistic legacy, it has often seemed to be the only plausible epistemological stopping place, and not unreasonably so.

This said, it needs also to be acknowledged that "skeptic" is an ambiguous term in its iterations across diverse circumstances. In present-day Western/Northern societies, tobacco use is a paradigmatically clear, even simple example, where a—frequently tacit—conviction prevails that its dangers have been established, whether or not people choose to act accordingly. Skepticism manifests widely and diversely in this and analogous matters. For purposes of this discussion, it arises urgently in debates surrounding the (putative or otherwise) dangers of climate change, especially in the variability of denials and contestations that permeate the discourse, and in the intensity of denials and oppositions. Denials range from simple (or not so simple) refusals to countenance assertions that there could *be* any issue, any cause for alarm, to affirmations of true urgency, evident in contentions that there is no time to waste: it is happening now. Frustratingly, evidence for or against can be mustered and/or contested across a range that runs from fear and urgency to skepticism often reinforced in "demonstrations"—usually experiential—of certainty to declarations that there is no cause for alarm. Setting the issues out in this way is by no means a declaration of sheer, bald incredulity. It is, rather, an attempt to point toward the multiple potential readings of "absolute certainty," which, oxymoronically, it would appear, are meant to bring disputation to an end, to establish the "truth." As to where such truths could reside, again disagreements persist. Presumably, there are good scientific and experiential readings on "either side." Were there not, these debates would long ago have been settled. But since the situation itself has clearly *not* been settled despite seemingly endless investigations and contestations, it might well be (humanly) impossible for *the truth* to emerge—even if an expectation of so definitive a conclusion is reasonable, despite the open-ended "nature" of the putative subject of inquiry. Hence, policy development is closely tied to something akin to a specific "world view" that supports or contests certain lines of action and inquiry on a local and a (putatively) global scale.

Climate change skepticism itself is multiply ambiguous and diverse in its iterations. Oreskes and Conway, for example, suggest that many such skeptics are committed to generating incredulity, thereby animating and justifying fervent resistance to measures designed/determined to regulate consumption across a range of practices, populations, and policies, often for reasons of local or personal social gain.[4] Some are skeptical about the very possibility of consensus on these issues, and for diverse reasons, not all of them commendable. Others are skeptical because they declare the current IPCC climate change position to be too gentle—insufficiently urgent—contending that it underestimates the severity of the threat climate change currently poses.[5] Doubts may be generated by a breed of epistemic distrust prompted by the concerns of ordinary folk about the difficulty of determining whom to trust, not just at an amateur, everyday level but globally, in situations where these "ordinary folk" are reliant on the expertise of others, whose good faith has to be taken on trust. In hesitations prompted by such diverse iterations of doubt, it is rarely easy to claim certainty, whether in thought or in action.

Puzzling in a rather different, if equally compelling, way are issues of certainty and/or uncertainty as these have been central to epistemological

---

4. Oreskes and Conway, *Merchants of Doubt*, esp. pp. 10–35.

5. My thanks to an anonymous reader of this manuscript for comments on these thoughts.

deliberation in Anglophone philosophy, at least since Ludwig Wittgenstein engaged them provocatively in *On Certainty*.[6] Skeptics commonly appeal to a margin of uncertainty to uphold *and* justify reluctance to declare certainty in knowing, whether individual or "collective," for some specified collectivity. Frequently, inquirers/would-be knowers withhold assent not merely out of perversity, but out of a justifiable wariness of exceeding the (perhaps yet unknown) boundaries of knowledge/truth. That said, a still more earnest reason for caution is—simply or not so simply—about exceeding the limits of the knowable, hence about going forward on shifting ground. Here, methodological critique may also claim a place in focusing on the—highly plausible—worry that the evidence accepted will almost certainly be partial, in both senses of the word: hence that it may be impossible ever to determine "the truth." And yet, the truth—"the conclusion"—will most likely be partial, ephemeral even though it would be careless to conclude that, therefore, ongoing inquiry is futile. Such a compromise would be unsatisfactory for numerous inquirers and in multiple epistemic situations. But for purposes of this project, now, I am suggesting that inference to the best explanation, however overworked it is as a contention, is often the most plausible—if interim—conclusion. It offers a way of "going on" and resists dogmatic closure. So even with climate change, which is clearly in process now, at the time of writing, it would still be implausible to declare certainty. So climate change, in a curious way, serves as a quasi-paradigmatic epistemic conundrum: How will "we" who are in the midst of it *know* when/if we really know and understand it, at least well enough to act wisely?

That said, it is vital to note—and try to understand—that climate change is a more complex, multiply contested, diversely manifested and experienced phenomenon than these too-brief suggestions imply. Claims about its causes and dangers vary widely, to the point where the very plethora of such challenges works to truncate aspirations for arriving at reasonable, collective, action-promoting conclusions. There may seem to be good reasons to reserve judgment in light of the clear inadequacy of such responses: reasons to postpone the inquiries for another day. Yet such delays would leave these issues untouched, and troubling in their ongoing persistence. Hence, I am contending, there are urgent, persistent—if multiply, locally diverse—*epistemic* responsibilities, both

---

6. Ludwig Wittgenstein, *On Certainty*, ed. G. E. M. Anscombe and G. H. von Wright, trans. Denis Paul and G. E. M. Anscombe (Oxford: Blackwell, 1968).

collective and individual throughout present-day societies, that urge sustained, wide-ranging ecological inquiry across a broad range of local and global living.

In the early days of twentieth-century white Western *feminist* ecological philosophy, with Val Plumwood eminent among its pioneering thinkers, questions about subjectivity were much in evidence, especially with respect to tacitly patriarchal colonizing theories and practices shaped by a curiously clumsy set of ontological connections between Woman generically conceived, and Nature. A feminization of "caring about nature," and not a laudatory one, was a strong contender in fostering such an orientation. So for Plumwood, "the rationalist tradition" in philosophy has been "inimical to both women and nature," especially in the connections it fosters between the "human self" and a range of instrumentalist practices.[7] To such aversions she attributes the conceptual aridity of pre-1990 environmental philosophy. On this issue, it is worth quoting her at length: Plumwood is justly critical of an account of ethics—and, by extension, of the "self"—in which "reason and emotion are sharply . . . opposed . . . and 'desire, caring, and love' are regarded as merely 'personal' and 'particular' as opposed to the putative universality and impartiality of [rational] understanding." Gendered alignments of reason with masculinity and emotion with femininity are unmistakable here. Likewise, traditionally "feminine" emotions are cast as untrustworthy, in contrast with "a superior, disinterested (and . . . masculine) reason."[8] For Plumwood, the task of what I call *ecological thinking* and practice is that of "reconceptualizing the human and reconceptualizing the self."[9] The project requires an extended critique of "the egoistic self of liberal individualism" and a turning away from instrumentalism, in moving toward viable-in-the-long-term conceptions of selves-in-relationship—more precisely, in positive, mutually respectful relationships, be they cordial or contestatory. Likewise, Plumwood sees instrumentalism as "a way of relating to the world which corresponds to a certain model of selfhood . . . conceived as that of the *individual* who stands apart from an alien other and denies his own relationship to and dependency upon

---

7. Val Plumwood, *Feminism and the Mastery of Nature* (London: Routledge, 1993), p. 142.
8. Plumwood, *Feminism*, p. 3.
9. Plumwood, *Feminism*, p. 5.

this other."[10] Entitlement is a predominant motif. Such a stance is so routinely taken for granted in affluent white Western-Northern societies that "we" have to learn to see it for what it is and to understand the presumptuousness of its deep-seated, tacitly endorsed ontology.

The shift Plumwood envisages here is no mere change in superficial everyday epistemic habits, nor does it entail transferring allegiance from one group, club, or political forum to another. It is more in the nature of an extended collective-cooperative conversion and learning process, performed gradually, perhaps thoughtfully—often angrily—with persistent, sustained effort. Thus, while no argument is required to confirm the aridity of a theoretical stance for which solitary, instrumental man is the guiding character ideal, its traces silently affirm the presence of *a presupposed thinking-acting self* who may seem to be race- and gender-neutral but whose presumptive maleness or even (*per impossibile*) gender neutrality tacitly perpetuates the constitutive effects of default andro-centered postures and values in and in relation to fellow living beings and to the wider natural world. Nor is such instrumental masculinity a characteristic of all men, but in the Western philosophical tradition, it refers—if tacitly—to (presumptively) white men: to the man of reason whose emblematic philosophical and social-political status Genevieve Lloyd productively deconstructs in her 1983 landmark text, *The Man of Reason: "Male" and "Female" in Western Philosophy*.[11]

By contrast, my thinking here takes its point of departure from the following manifesto with which I conclude the introduction to *Ecological Thinking*:

> Thinking ecologically carries with it a large measure of responsibility—to know somehow more *carefully* than single surface readings can allow. It might seem difficult to imagine how it could translate into wider issues of citizenship and politics, but the answer, at once simple and profound, is that ecological thinking is about imagining, crafting, articulating, and endeavouring to enact principles of ideal cohabitation.[12]

---

10. Plumwood, *Feminism*, p. 18.

11. Genevieve Lloyd, *The Man of Reason: "Male" and "Female" in Western Philosophy*, 2nd ed. (London: Routledge, 1994). In the twenty-first century, one might suggest that the ideal has been displaced by or devolved into that of *homo oeconomicus*.

12. Lorraine Code, *Ecological Thinking: The Politics of Epistemic Location* (New York: Oxford University Press, 2006), p. 24.

These thoughts bring social epistemology and ethical-political theory together in relation to matters ecological, and to Anglophone feminist and antiracist theories of the late twentieth- and early twenty-first centuries: they pick up the threads at places where these lines of inquiry intersect. This conceptual frame extends into matters of vulnerability, incredulity, ignorance, and trust: modalities unevenly distributed across populations of internal diversity and of diverse relations to/with the rest of the world. Hence social justice issues—which are at the core of this inquiry—cannot be investigated in one-size-fits-all analyses: they have to strike a balance across human, other-than-human, and situational particularity, multiplicity, and commonality. They lead into questions about education, human rights and the environment, the politics of care in techno-science, and the public cultivation of ignorance and doubt in climate change skepticism, that foster an atmosphere of incredulity and distrust.

These many years later, such conclusions are less plausible than they could have been then. The ideals they affirm are increasingly elusive in times of often brutal scarcity in a time when identity politics—ontologically and materially—contests simple assumptions about human sameness and equality of aspiration. Tacit egalitarian expectations crafted in the affluent early Western-Northern "modern" world continually, and brutally, come up against their own practical-ontological expectations to expose the limited reach of such aspirations. Thoughts such as these pose fundamental, challenging questions about who we think we are; we who have made the world for which we are now accountable, who need to craft ways of restoring, renewing, preserving that world for others who come after us and/or live alongside us; other generations, genders, cultures, races, and species who must live with the effects of our practices. I want, therefore, to urge speaking and hearing the question "Who do we think we are?" provocatively, normatively, not as a rhetorical question, but as an outraged political-ontological challenge posed in astonished dismay at presumptuous behaviors, intrusive or offensive assumptions, fantastic-extraordinary achievements. Posing it in this way invokes social justice as both a fundamental and an overarching ethical-epistemological-political concern, spoken from positions shaped by feminist, postcolonial, and other Others' engagements in recognizing that knowing responsibly (whether singly or collectively) is *sine qua non* for responsible social action. The addressive formulation emphasizes the question's necessarily interactive character. It contests tacitly or overtly individualistic ontological presuppositions and practices.

These thoughts pertain principally to Anglo-American analytic philosophical orthodoxy where "individuals" are starkly "individual" (if rarely individuated) and, in consequence, supremely self-interested. A principal goal of thinking ecologically, therefore, is to unsettle an epistemological orthodoxy that upholds widespread white Western/Northern ontological convictions about discretely self-contained human individuality, citizenship, and homogeneity as these underpin "our" presumptive entitlement to live, consume—and pollute—as we will. This goal prompts "us," as educators and activists, to engage in determining how to generate radical shifts in going assumptions about "the ontology of the self," in policies and practices that govern quotidian public-political social structures in the affluent white Western world. One vital piece would be to instill a measure of caring and of humility—of affect—that, Plumwood notes, is too often held at a distance from *bona fide* questions about knowing, or polluting, even though it is germane to engaging with them well. This claim is one of the hardest to defend and the most vital to enact.[13]

A principal contrast is with practices of superimposing categories and explanations onto events, situations, and "people" that subsume differences under preformed categories, with inadequate attention to whether an appropriate "fit" is achieved/achievable. It is perhaps an inadvertent practice, yet one that is endemic in background ontological assumptions that inform the epistemologies and the ethical-political theories and practices of the current white Anglo-American mainstream. Hence, in George Yancy's aptly titled volume *The Center Must Not Hold*,[14] a vital catalyst is this very question: Who do "we" think we are in taking for granted our entitlement to live as we do? It is implicit, to cite a small sampling of examples: in Susan Babbitt's analysis of the rarely noticed "whiteness" of the philosophy practiced in most English-speaking universities, with its multiple exclusionary effects; in Shannon Sullivan's thoughts about how present-day Anglo-American philosophy's upheld secularity, particularly

---

13. Consider Mick Smith's critique of "the ethical" in modern society, where he writes: "The heartfelt aspects of ethics—of love, care, and sympathy—are turned into social currency to facilitate trade-offs between people with different interests and aims." Mick Smith, *An Ethics of Place: Radical Ecology, Postmodernity, and Social Theory* (Albany: SUNY Press, 2001), p. 65.

14. George Yancy (ed.), *The Center Must Not Hold: White Women Philosophers on the Whiteness of Philosophy* (Lanham, Md: Lexington Press, 2010).

in the United States, works to discourage people of color from participating in the discipline because "[I]n many philosophical circles, to be a person of faith is to be perceived as ignorant, backward, and primitive"; in Alexis Shotwell's explorations of the effects of "whiteness as method" in sustaining a wide range of exclusions and oppressions.

Borrowing a conceptual framing from Robert M. Figueroa and Gordon Waitt, I propose that openness to the potential *epistemic* legitimacy of "affect, materiality, creativity, and multiplicity" creates spaces for capacious ways of recognizing ecological subjectivity, where "[n]ormative elements of moral terrains include embodied narratives about what behaviours are permissible, who belongs where, how we perceive the moral status of other bodies (human and non-human), and the ways in which we establish moral, social, and political identities in embodied relations to space and place."[15] Their reference is to indigenous Australians' requests, addressing would-be climbers, that they respect the sacredness of Uluru in ways that require them not just to *know* differently, but to *be* different from how they have assumed they can be in regarding the natural world as up for grabs—or for climbing!—at anyone's whim. It requires endeavoring to *know* Uluru not by the colonizers' label "Ayers Rock," with all this possessive implies, but as a sacred site that is not theirs (= the climbers') to appropriate, and where climbing counts as appropriation: it calls for a reconceived ontology, and an ethics of place sufficiently sensitive to *recognize* specificities and particularities in ways that respect and honor them. It calls upon people to rethink their (often hyper-individualized) sense of self, of place, and of ownership, to move away from convictions of autonomous entitlement while working to achieve respect for an otherness that is not theirs to obliterate or to own, and whose standing is not theirs to define or to dispute. Citing just one striking example, the "Who do you think you are?" question could be posed challengingly to the young white man who insists, disregarding explicit aboriginal requests about respecting Uluru, "I climbed it anyway. But, that's just the way I am. I'm not going to not climb it."[16] *Who* does he think he is?

---

15. Robert M. Figueroa and Gordon Waitt, "Climb: Restorative Justice, Environmental Heritage, and the Moral Terrains of Uluru-Kata National Park," *Environmental Philosophy* 7 no. 2 (Fall 2010), p. 147. A ban on the climb of Uluru took effect on October 26, 2019.
16. Figueroa and Waitt, "Climb," p. 151.

How these thoughts bear on larger ecological-epistemic projects of developing principles of ideal cohabitation might be puzzling. But working away from the radical individualism and the universal entitlement endemic in white Western Anglo-American philosophy and social theory, endeavoring to think and act empathically, communally, co-operatively across human and situational differences is germane to responding fittingly to ecological harms and/or benefits in their specificity and generality. Their impacts cannot responsibly be known or enacted in one-size-fits-all approaches. Contesting the tenacity of the center simultaneously contests the applicability and pertinence of situation-insensitive policies and of remedial measures elsewhere, in their diversity and particularity.

Engaging knowledgeably in/with circumstances such as these is ontologically challenging for it requires rethinking, re-enacting *who we are* in ways sufficiently sensitive to dislodge such entrenched convictions. This is the hardest, most urgent requirement: it is easier, more imaginable, to think and participate on the surface, so to speak, in revisionary ways of doing, thinking, knowing. But to practice a philosophy and develop a pedagogical practice which requires—indeed, which *must* require—such rethinking and reenacting is ontologically-affectively-epistemologically unsettling; it has to be a gradual, careful process. Requirements such as those I have noted are distantly analogous to late twentieth-century consciousness-raising practices: they call for radical critical-genealogical analyses of who we are and how we are accustomed, unthinkingly, to live. Such analyses might well unsettle many of the expectations that inform "our" everyday thoughts and actions: such is their (justifiable) power. Although their work is by no means complete, the effects of feminist, postcolonial, and antiracist consciousness-raising projects and practices have been far-reaching in ways that are exemplary for unsettling these and analogous ecological practices.

Without doubt, as I have noted, so fundamental a requirement is difficult to fulfill: it is easier to devise revisionary, surface ways of doing, thinking, knowing. But to practice a philosophy that requires unsettling and reenacting—"making strange"—long-standing assumptions about *who we are* is ontologically and epistemologically radical, upheld as these assumptions are by an instituted social-political-epistemic imaginary (following Cornelius Castoriadis)[17] in which we inhabitants of the affluent

---

17. Cornelius Castoriadis, "Radical Imagination and the Social Instituting Imaginary," in Gillian Robinson and John Rundell, eds., *Rethinking Imagination: Culture and*

Western-Northern world live and think, however obliquely or contrarily. In Castoriadis's analyses, the social imaginary is neither static nor fixed. Its allegiance to an Enlightenment conception of reason in which a transcendent knowing subject dispassionately contemplates and investigates the world as *he* finds it is held in place by what Castoriadis calls an *instituted social imaginary*: "a *world* of social imaginary significations whose instauration as well as incredible *coherence* goes unimaginably beyond everything that 'one or many individuals' could ever produce."[18] To it he opposes the idea, and the energy, of an *instituting* imaginary through which imaginatively initiated counterpossibilities interrogate the going social structures, radically and persistently, to destabilize their pretensions to "naturalness" and "wholeness," thereby initiating a new making. Castoriadis's interest is in how the imaginary of (Western/Northern) late capitalism produces and sustains unjust social hierarchies, perpetuates a mythology of the instrumental innocence and neutral expertise of scientific knowledge, and generates illusions of benign connections between power and knowledge: in how its assumptions about relations between "individuals" and societies work to create and legitimate structures of domination and exploitation. An *instituting* imaginary, by contrast, is a vehicle of radical social critique. It is about exposing, learning to understand, and working to reconfigure the power-infused rhetorical spaces where knowledge making and circulating occur. The larger vision is global, but the activities—the praxes—ordinarily occur locally. Such contestations grow out of what Castoriadis applauds as the critical-creative activity of a society or a social group that manifests in its capacity to put itself in question, prompted by a recognition that—as a society—it is incongruous with itself, with scant reason for self-satisfaction.[19] (There are historical precedents, of which Voltaire's *Candide* counts among the best-known, still-pertinent examples.[20]) In this regard, educators' epistemic responsibilities need to include constructive, affective engagements with matters ecological. This vital knowledge has to inform our pedagogical practices.

Creativity (London: Routledge, 1994). (See my references to Castoriadis in *Ecological Thinking*, pp. 123–25.)

18. Cornelius Castoriadis, *Philosophy, Politics, Autonomy: Essays in Political Philosophy*, ed. David Ames (New York: Oxford University Press, 1991), p. 62, emphasis in original.

19. See my discussion of Castoriadis in *Ecological Thinking*, p. 195.

20. Voltaire (1694–1778), *Candide*, trans. Daniel Gordon (Boston: Bedford St Martin's, 1999).

Such a responsibility could/should be realizable, also, in generating an intelligent skepticism in our students, colleagues, and other interlocutors about the taken-for-granted rightness of who "we" are and how "we" live. Responsibility-in-action of this nature is writ large, for example, in the activities of such early twenty-first-century political protest groups as the Occupy movements.

In her 2001 book *Retrieving Experience*, Sonia Kruks suggests: "[T]oo strong an identification with others permits us to deny the responsibilities . . . *born of our own location*."[21] I cite this claim on the way to articulating a "healthy skepticism" about the reach of the aggregated first-person plural pronoun "we" and about the extent and the quality of "our" epistemic and affective capacities to understand one another, whether "personally" or "situationally," across stark otherness. Yet I mean also to affirm that, in putatively "free" societies, human lives are, if variously, structured by responsibilities to endeavor to do just that, if never perfectly to achieve it. Although things, lives, situations, and experiences are neither so homogeneous nor so commensurable as blithe appeals to an unmarked "we" assume, neither is radical incommensurability a tenable assumption, especially when/if it is invoked as an "excusing," "exonerating" factor to justify epistemic or moral-political inertia or consciously sustained modalities of ignorance, of unknowing in hierarchically structured societies and social-political movements.

It is vital, then, to the achievement and maintenance of social justice broadly conceived, for would-be knowers to work, collaboratively and singly, toward developing an apt measure of *hermeneutic* understanding with respect to negotiating knowledgeably and empathically with one another across diverse situations, "identities," and circumstances. I invoke hermeneutics to affirm that this "imperative" is vital, yet not starkly cognitive in a positivist-derivative empirical sense: such cognition, conceived positivistically, would carry neither the capacity nor the license to address/interpret/understand the multiple *affective* dimensions of being human, in concert with or apart from the cognitive dimensions of lives situated outside the tacit norms of epistemic sameness (by which I mean, to differing degrees, most lives). Thus, Sandra Bartky is appropriately wary of the persuasive force of assumptions that *knowing*, properly conceived, reduces to acquiring information. She advocates working

---

21. Sonia Kruks, *Retrieving Experience: Subjectivity and Recognition in Feminist Politics* (Ithaca: Cornell University Press, 2001), p. 158 (emphasis added).

toward "a knowing that transforms the self who knows, a knowing that brings into being new sympathies, new affects as well as new cognitions and new forms of subjectivity . . . in a word . . . a knowing that has a particular affective taste."[22] Such ways of knowing are continuous with those intended to inform ecologically oriented consciousness-raising and the social-political activities it generates. There is no infallible true-false mechanism for ticking off the putative items of knowledge that "relational" knowing of these kinds requires or can achieve: the process may be fraught, even dangerous in its capacity to distort, intrude, and/or enact damage.[23] Here one might locate the source of empiricist-analytic resistance to acknowledging the epistemic value of such affect-infused inquiry, especially for the—admittedly unquantifiable—understandings it enables. Yet, affect in its multiple modalities is integral, and indispensably so, to thinking and acting well ecologically, interactively in its literal and its derivative, quasi-metaphorical senses.[24]

At its best, the attentive, temporally extended, process-observant, *listening* dimensions of Bartky's proposal signal radical—and epistemologically worthy—departures from the punctiform, one-off, dislocated knowing articulated in true-or-false propositional form, of standard late-twentieth-century Anglo-American epistemology, with its veneration of a presumptively individual "view from nowhere." A further example amplifies this claim: Bronwyn Hayward (referring to Hannah Arendt) argues for what I perceive as the ecological-epistemological significance of good, attentive listening—across diverse experiences, understandings, and/or "takes" on specific situations and events. Addressing the complexities of developing strategies for rebuilding a city intelligently, in Christchurch, New Zealand, following the 2010 earthquake, Hayward notes: "[M]aking room for dialogue and listening seems a prudent strategy if we wish to develop an idea of 'determinative morality,' or a vision of what we

---

22. Sandra Bartky, "Sympathy and Solidarity: On a Tightrope with Scheler," in Diana Meyers, ed., *Feminists Rethink the Self* (Boulder, CO: Westview, 1997), pp. 178–79. I discuss Bartky's view in *Ecological Thinking*, chapter 6.

23. See my "'I Know Just How You Feel': Empathy and the Problem of Epistemic Authority," in Lorraine Code, *Rhetorical Spaces: Essays on (Gendered) Locations* (New York: Routledge, 1995), pp. 120–43.

24. Pertinent is Clare Hemmings's, "Affective Solidarity: Feminist Reflexivity and Political Transformation," *Feminist Theory* 13 no. 2 (Aug 2012), pp. 147–61, to which I return in chapter 5.

ought to do, not just what we ought *not* to do."[25] Much of mainstream Anglo-American epistemology, with its constitutive individualism and concomitant emphasis on propositional verifiability, claims scant conceptual resources for evaluating such unquantifiable/unverifiable cognitive practices as listening, disputation, and dialogue: it is impoverished in consequence. Hayward's comment points toward an urgent need for communal practices of engaging, intelligently and affectively, in (often extended) listening practices, in complex social-political situations where simplistic, one-size-fits-all propositions modeled on the "S knows that p" structure are inadequate to the task of understanding well enough to enable acting well, generically, so to speak. Helpfully instructive is Donna Haraway's 1997 plea for a critically engaged politics of science-knowledge, where by "critical" she intends practices of inquiry that are "evaluative, public, multiactor, multiagenda, oriented to equality and heterogeneous well-being."[26] There she applauds a then-Danish "practice of establishing panels of ordinary citizens, selected from pools of people who indicate an interest, but not professional expertise or a commercial or organized stake, in an area of [science and/or] technology" who engage in temporally extended processes of hearing testimony, reading briefs, deliberating among themselves, and reporting to a wider public.[27] Such processes foster what she has called "situated knowledges." I reiterate this view for its consonance with the practices Hayward gestures toward in the deliberative strategy she advocates and for its capacity to dislodge the emblematic individualism, for ecological inquiry, of the "pure research" conducted in the ivory towers, from which Haraway distances her analyses.[28] The point is not, implausibly, to discredit factual knowledge *tout court*, either in general or for ecological purposes, but to be ever mindful of its scope and limits.

---

25. Bronwyn Hayward, "The Social Handprint: Decentering the Politics of Sustainability after an Urban Disaster" in Peg Rawes, ed., *Relational Architectural Ecologies: Architecture, Nature and Subjectivity* (London: Routledge 2013), p. 240.

26. Donna Haraway, *Modest_Witness@Second_Millenium. FemaleMan Meets_OncoMouse* (New York: Routledge, 1997), p. 95.

27. Haraway, *Modest_Witness*, p. 96.

28. In like vein, Kristin Shrader-Frechette advocates deliberative democracy as a civic virtue. In her view (indebted to Iris Marion Young), it "demands commitment to universal justice, to listening to others . . . to avoiding blind rage, violence, and nihilism" (my italics). *Taking Action, Saving Lives: Our Duties to Protect Environmental and Public Health* (New York: Oxford University Press, 2007), p. 10.

Space does not permit a fuller articulation of epistemic resources beyond the confines of analytic epistemology, but there are multiple such resources in so-called "continental" philosophy, especially in its hermeneutic iterations. They tend not to announce themselves as epistemological—perhaps from a resistance to the formality of the label itself—but their effects are frequently to enrich cognition beyond the confines of propositional knowing, and specifically, to engage *listening* as a central modality of epistemic practice. In an example of how listening attests to a fundamental human capacity to open oneself to the world, also—if differently—consonant with Bartky's resistance to the "knowledge-as-information" model, Martin Heidegger writes: "[L]istening to . . . is Dasein's existential way of Being-open as Being-with for Others."[29] The idea of listening as fundamental to being human and as ontologically constitutive of social being (*Mitsein*) points to the artificiality and the limitations of models of knowledge reliant on discrete input-output fact finding. Hence, with regard to the guiding question "Who do we think we are?" it would be implausible to respond that we are creatures for whom *seeing* isolated medium-sized physical objects (such as cats on mats), isolated from their surroundings, is believing/knowing; nor can such stark individualism serve ecological purposes well.[30]

This evocation of Heidegger is intended to claim him, unequivocally, neither as an ecological thinker nor as a protofeminist,[31] but to

---

29. Martin Heidegger, *Being and Time*, trans. John Macquarrie and Edward Robinson (New York: Harper & Row, 1962), p. 206. "Dasein" is Heidegger's term for human being. Literally, it translates as "being there" and is evocative for discussions such as this in its affirmation that human lives are always situated, somewhere, there. A useful resource (among many) in this regard is Krzysztof Ziarek, *Inflected Language: Toward a Hermeneutics of Nearness: Heidegger, Levinas, Stevens, Celan* (Albany: SUNY Press, 1994).

30. For further thoughts on such matters, see Steven Vogel, "Nature as Origin and Difference: On Environmental Philosophy and Continental Thought." *Philosophy Today*, SPEP Supplement, 1998, pp. 169–81.

31. Here I am indebted to Trish Glazebrook's analysis in "Heidegger and Ecofeminism," in Nancy J. Holland and Patricia Huntington, eds., *Feminist Interpretations of Martin Heidegger* (University Park: The Pennsylvania State University Press, 2001), pp. 221–51. In "Introduction 1—General Background," Huntington notes that for Heidegger: "All understanding occurs on the basis of a mode of affective attunement that colors our perception and the overall way in which the world appears intelligible to us" (p. 10).

draw attention to his hitherto rare (in Western-Northern epistemology) emphasis on *listening* as ontologically-affectively fundamental to human being *tout court*, by contrast with the primacy accorded, in dominant Anglo-American epistemic discourse, to on-off propositional knowledge claims reporting visual or tactual "givens," uttered as though to everyone and, in effect, to no one. It is to emphasize the *impotence* of such affect-free approaches to generate viable understandings of complex social-political-ecological situations.

So, for example, Trish Glazebrook convincingly observes:

> [T]he truth of modern science, and the truth of technology are ways in which human being knows nature. But *Denken* and *Besinnung* are thoughtful, respectful, and thankful relations to nature, rather than its reduction to object and resource. Heidegger's vision is an ethic of reciprocity and care, the very vision for which ecofeminists call that stands in marked contrast to what has been diagnosed and rejected as a logic of domination by both.[32]

These ways figure among the relational modalities listening invokes: they are antithetical, and helpfully so, to individualistic/positivistic ways of knowing; nor, strikingly, do they presuppose or initiate an artificial separation between ethics and epistemology. Such thoughts are integral to analyses and practices of what I have elsewhere called epistemic responsibility.[33] They may not be ubiquitously appropriate or epistemologically productive, but they attest to a way of thinking for which the world/the earth does not reduce to (Heideggerian) "standing reserve," a term that, for Heidegger, refers to regarding the world as merely an energy resource, a thing to be used. Putting the point strongly, Heidegger writes: "[E]verywhere everything is ordered to stand by, to be immediately on hand, indeed to stand there just so that it may be on call for a further ordering. We call it the standing-reserve [*Bestand*]."[34] Further, he cau-

---

32. Glazebrook, "Heidegger and Ecofeminism," p. 243.

33. Lorraine Code, *Epistemic Responsibility* (Hanover, NH: University Press of New England, 1987).

34. Martin Heidegger, "The Question concerning Technology," trans. William Lovitt, in David Farrell Krell, ed., *Martin Heidegger: Basic Writings* (New York: Harper & Row, 1977), p. 298.

tions: "[A]s soon as what is unconcealed no longer concerns man even as object, but exclusively as standing-reserve, and man . . . is nothing but the orderer of the standing-reserve . . . then he comes to the point where he himself will have to be taken as standing-reserve."[35] Hence, with reference to the basic ontological question "Who do we think we are?" a principal worry must be that a reduction of the world to standing-reserve would threaten to reduce humanity—human being—to the status of "bare life."[36] As the "who" question suggests, this complex problematic reaches out, rather, for engagement in developing something akin to what Michel Foucault has called "a critical ontology of ourselves"—here focused on the "we" in "we-saying."[37]

How, then, do *these* thoughts bear on "we-saying"? Along the way to responding to this question, I propose recalling Robert Reich's February 2014 posting to *Nation of Change*. Under the heading "America's 'We' Problem," he aptly observes:

> [T]he pronouns "we" and "they" are the most important of all political words. They demarcate who's within the sphere of mutual responsibility, and who's not. Someone within that sphere who's needy is one of "us"—an extension of our family, friends, community, tribe—and deserving of help. But needy people outside that sphere are "them," presumed undeserving unless proved otherwise.'[38]

This, in brief, is America's "we" problem, at least with respect to wide-ranging social-political situations and circumstances, if not precisely to matters ecological. It is often articulated in such seemingly simple questions as "Why should we pay for them?," incredulously posed in affluent twenty-first-century societies in resistance to buying health insurance that will

---

35. Heidegger, "The Question," p. 308.
36. Mick Smith, *Against Ecological Sovereignty: Ethics, Biopolitics, and Saving the Natural World* (Minneapolis: University of Minnesota Press, 2011), p. 106. "Bare life" is from Giorgio Agamben, *Homo Sacer: Sovereign Power and Bare Life*, trans., Heller-Roazen (Stanford: Stanford University Press, 1998).
37. Michel Foucault, "What Is Enlightenment?" trans. Catherine Porter in Paul Rabinow, ed., *The Foucault Reader* (New York: Pantheon Books, 1984), p. 47.
38. See <http://www.nationofchange.org/america-s-we-problem-1392478228#comments>. Last accessed 12 February 2015.

help to pay for less affluent people with preexisting health problems or to extending employment benefits to the long-term unemployed.

While Reich reasonably dubs the problem "America's," its extension through affluent parts of the Western-Northern world should not be overlooked. It is frequently posed by the environmentally ecologically "well situated" to justify excessive water usage, food or fuel consumption, or air pollution with the argument that "we" who have earned the right to do so can claim a certain entitlement; other less well-situated people have not, and cannot. The implication is that the "haves" have right on their side, that they count as the deserving "we," they have set the tone for over-consumption by affluent populations throughout the post-World War II Western-Northern world and beyond. It speaks to a slanted, distorted nod in the direction of an unjust version of social justice. These are the unselfconscious "we-sayers." By contrast, Adrienne Rich, in her landmark feminist challenge "Notes for a Politics of Location," raises the matter of we-saying self-consciously and insistently. Observing that radical feminists had sought nothing less that "the making new of all relationships," she poignantly observes: "The problem was that we did not know whom we meant when we said 'we.'"[39]

The topic of "we-saying" is large and amorphous beyond the range of this introductory discussion, but becoming attuned to its effects is vital in elaborations of my claim that ecological inquiry tends to begin "too far up" on an imagined line, in the sense that it often fails to address these most fundamental questions. These are ontological questions about who we think we are and how we live the "answers." Failure to acknowledge their political salience is widespread in debates about ongoing ecological crises. It would be impossible to address them adequately here. Yet they are urgent questions precisely because of their too-frequent invisibility and because of the startle effect sparked by Reich's reminder of the unself-conscious ways of referring to what "we" think, do, believe, and want that permeate white Western everyday and academic talk.

Stephen Gardiner poses such questions challengingly when he asks, "[A]re we the scum of the earth?"[40] insisting that ethics is not just about

---

39. Adrienne Rich, *Blood, Bread, and Poetry: Selected Prose 1979–1985* (New York: Norton, 1994), p. 217.
40. Stephen M. Gardiner, "Are We the Scum of the Earth? Climate Change, Geoengineering, and Humanity's Challenge," in Allen Thompson and Jeremy Bendik-Keymer, eds., *Ethical Adaptation to Climate Change: Human Virtues of the*

what "we" should *do*, but about "who we are." His challenges are articulated within a framework of philosophical ethics, but they presuppose no artificial separation between epistemology and ethics, whose reciprocal influence is apparent, if not explicitly stated. The question is both urgent and troubling. There is a certain irony in the tacit assumption that we who ask it *know* who "we" are: both the "we" who pose it, and the "we" to whom it is posed. So, endeavoring to answer the question is at once compelling and restricted from the outset by its own terminology and the assumptions silently embedded in them. The point is emphatically not to avoid "we-saying" in such situations, but to be ever mindful of its multidimensional self-referentiality in ways that point toward ongoing reconfigurations of *praxes* rather than toward a need for discrete factual or moral "corrections." With this reference to praxis/praxes, I again signal the reading of "we" that informs this book. It is vital to spell out its implications and challenges so as *explicitly*—thus not merely tacitly or presumptively—to insist on the significance, for ecological discourse, of resisting appeals only or primarily to "individual" action and condemnation and of allegedly self-announcing "one-liners." Hence, in invoking praxis, I am, again loosely, indebted to Hannah Arendt's introduction of its Aristotelian source where a human life is "always full of events which ultimately can be told as a story, establish a biography; it is of this life, *bios* as distinguished from mere *zoe*, that Aristotle said it is 'somehow a kind of *praxis*.'"[41] Invoking *praxis* enables Arendt to link action to freedom and plurality, to view action as "a mode of human togetherness" that stands in sharp contrast to bureaucratic, elite modalities of political action such as those that characterized modernity as she knew it.[42]

Praxes such as those toward which Arendt gestures, and practices more generally pivotal to teaching, acting, and endeavoring to generate epistemically responsible ways of approaching life the universe and everything (with a nod to Douglas Adams[43]), require recognition of the

---

Future (Cambridge, MA: MIT Press, 2012), p. 242 (italics original).

41. Hannah Arendt, *The Human Condition* (Chicago: University of Chicago Press, 1958), p. 97. The Aristotle quotation is from the *Politics* 1254a7.

42. I am drawing on Maurizio Passerin d'Entreves, "Hannah Arendt," in *The Stanford Encyclopedia of Philosophy* (Winter 2013 Edition), ed. Edward N. Zalta, <http://plato.stanford.edu/archives/win2013/entries/arendt/>. Last accessed 12 February 2015.

43. *Life, the Universe and Everything* (Pan Books, 1982) is the third volume in Douglas Adams's *Hitchhiker's Guide to the Galaxy* series.

extent to which education at every stage and in every facet of life, both commonly and individually, is fundamental to how "we" situate ourselves and act within the ecological circumstances that are everywhere around "us." Indeed, commitments to achieving epistemic responsibility require finding ways to answer Gardiner's question with a resounding "no!" which can be uttered and enacted honestly only when/if a measure of humility is achieved. A *just* measure of humility will attest to a recognition of the limits of human knowledge, however sophisticated and/or arcane the subject matter may be and however erudite the knower. It is against a background of such thoughts that William Throop names humility (understood as a "tendency to restrain self-interest, and sensitivity to the idiosyncrasies of a system and its surroundings"[44]) as a primary, pervasive ecological virtue that attests to respect for the earth and for other living beings. Justly achieved respect requires knowing well the entities with/for/against which action is undertaken; perhaps more significantly, it requires resisting the arrogance of undue self-promoting, standing back and finding no lessening of "self" in acknowledging places or areas of not-knowing, even of ignorance. In the view I am advancing here, such attributes are as collective as they are individual: a hard recognition to achieve for those trained to affirm epistemic individualism. It may well be that these attributes are commonly enacted individually, but in matters ecological it is vital, always, to be cognizant of their collective import, both for ways of being and knowing and for the earth.

Here again, Arendt contributes thoughtfully. She writes:

> [E]ducation is the point at which we decide whether we love the world enough to assume responsibility for it, and by the same token save it from that ruin which except for renewal, except for the coming of the new and the young, would be inevitable. And education, too, is where we decide whether we love our children enough not to expel them from our world and leave them to their own devices, nor to strike from their hands their chance of undertaking something new, something unforeseen by us, but to prepare them in advance for the task of renewing a common world.[45]

---

44. William M. Throop, "Environmental Virtues and the Aims of Restoration," in Thompson and Bendik-Keymer, eds., *Ethical Adaptation to Climate Change*, p. 55.

45. Hannah Arendt "The Crisis in Education," 1954, available at <http://www.hannaharendtcenter.org/?p=7983>. Accessed 27 February 2015.

It prepares them/us also for sustaining some version of what Shannon Sullivan, referring to what she learned in school, calls "ignorance/knowledge."[46] This thought returns the discussion to where it began: to asking who we think we are, reframed in relation to ways of world making. It offers some guiding ideas about how "we" might think/enact who we are and how to work toward being that way—ecologically, at least—or perhaps more completely, in the manner of "our" dwelling in/with the world.

---

46. Shannon Sullivan, "White Ignorance and Colonial Oppression: Or, Why I Know So Little about Puerto Rico," in Shannon Sullivan and Nancy Tuana, eds., *Race and Epistemologies of Ignorance* (Albany: SUNY Press, 2007), pp. 153–72.

Chapter 1

Epistemic Responsibility, Now

Borrowing its title from my 1987 book *Epistemic Responsibility*, this chapter's central purpose is to invite its readers to think again—now, in the twenty-first century—about what it means to work from a commitment to responsible epistemic conduct, both singly and collectively, as a regulative ideal in projects of seeking, achieving, contesting, and communicating knowledge. Such responsibilities attach both to "individual" knowers and to collective knowing—that is, to speakers, educators, writers, readers, thinkers across diverse spaces, options, inclusions, and exclusions of human social-political living. Commonly, responsibilities of this nature do not speak in monologue, but are (if silently) addressive—as much in their reliance on speaking to/with others, as in their intent and effects. Hence, in a Western-Northern epistemic era when the putative certainties integral to and consequent upon positivist inquiry seem no longer to hold, the epistemic environment is increasingly volatile. Entering that environment philosophically, engaging with its multiple implications, is likewise challenging, and compelling.

With respect to matters of concern in the "real" world, honoring these responsibilities requires and derives from commitments to acting knowingly, justly, understandingly and often humbly, in interactions with other people, with other species, and with/in the world/the earth. When they are well enacted, such commitments manifest engagement—*responsible* epistemic engagement—with interlocutors, and with the world both animate and inanimate, where "knowing responsibly" entails avoiding any precipitous "rush to judgement": it manifests in a readiness to think and rethink, in commitments to practicing the humility required for acknowledging, and for backing away from too-hasty judgement, misunderstanding, or error. Here, knowing is no solitary endeavor even in a minimal sense: it is an engaged, dialogic, collaborative/contestatory process and practice. In Western-Northern societies, such responsibilities are integral to realizing and promoting sociality in both a descriptive and a normative sense; these are vital ingredients for engaging thoughtfully, affectively, critically and/or constructively with knowing as a social-political, and often cooperative practice. They extend, especially in ecologically aware philosophical practices, from "individual" knowers and singular epistemic actions to sites of multiplicity, interaction, affirmation, contestation. Yet they have also to strike a mean: to be wary of too hasty conclusions, and still more wary of the risks of venturing too far, and enacting harm/damage in the process at issue. Such knowing has commonly to make space for engaging with "place" as integral to the production and circulation of knowledge—place as initially brought into Anglo-American thinking by Donna Haraway's ground-breaking introduction of "situated knowledges" into epistemic discourse.[1] "Situation" recognized, here, in being acknowledged, points toward working to achieve and communicate awareness of the invasive power of putative knowing, just as clearly as it points toward realizing and respecting the potential of untried epistemic experiments, explorations of new epistemic territory.

With a renewed constructive-critical emphasis on testimony in Western-Northern Anglophone theories of knowledge in the twenty-first century, then, comes a need for those of "us" who position ourselves as responsible knowers to rethink "who we are" in claiming to know. In such rethinking, matters of epistemic responsibility claim a renewed pertinence. Here, this chapter builds upon some of the central thoughts in my eponymous 1987 book—*Epistemic Responsibility*—to elaborate conceptions of epistemic subjectivity, responsibility, and agency that inform subsequent chapters in this book. It focuses on the "nature" and multiple, complex dimensions of living as a responsible knower in the twenty-first-century Western-Northern world: on particularity, relationality, testimony, and again on the "nature" of putative knower(s) and on their ways of occupying and enacting their positions as participants in an (epistemic) ecosystem.

---

1. Donna J. Haraway, "Situated Knowledges: The Science Question in Feminism and the Privilege of Partial Perspective," in *Simians, Cyborgs, and Women: The Reinvention of Nature* (New York: Routledge, 1991), pp. 183–202.

In her 2005 book, Adriana Cavarero observes: "Uniqueness is epistemologically inappropriate." This observation may strikes a discordant note for practitioners of Western-Northern analytic epistemology in its commitment to seeking necessary and sufficient conditions for establishing the universal validity of items of "knowledge in general." Cavarero contends that in establishing the overarching superiority of a putatively universal *logos*, "the philosophical tradition . . . ignores the unrepeatable singularity of each human being, the embodied uniqueness that distinguishes each one from every other."[2] I open the chapter with this observation to propose that engaging responsibly, epistemologically, and morally-politically with human specificity and vulnerability requires "us"—whoever we are—to counter any assumption that uniqueness, *particularity*, merits epistemic attention only to the extent that it informs and sustains the universal. Hence, endorsing Cavarero's contention as a working hypothesis, in this chapter, I consider some implications of an injunction against uniqueness as it has infused established Anglo-American epistemology, most notably in moral epistemology, but just as significantly in analogous and more widespread epistemic practices reliant on knowing people and circumstances well, in their singularity and multiplicity. Cavarero's apt charge that even "philosophies that value 'dialogue' and 'communication' remain imprisoned in a linguistic register that ignores the *relationality* already put in action by the simple reciprocal communication of voices" (16, my emphasis) enjoins caution with respect to any complacent imagining that the anonymous, monologic pronouncements of the white Western philosophical tradition have been displaced in favour of such relationality as she advocates. She is speaking from a disenchantment with the monological tenor of standard twentieth- and twenty-first-century propositional knowledge claims, spoken as though to everyone or no one: with their utterance into presumptively neutral spaces that accord little significance to the specificities of interlocutors, circumstances, and places that afford, or fail to afford, the conditions of their uptake and acknowledgment.

Engagement with particularity is a recurrent theme for Cavarero. It emerges eloquently in her earlier book, *Relating Narratives: Storytelling and Selfhood*, where she writes:

---

2. Adriana Cavarero, *For More than One Voice: Toward a Philosophy of Vocal Expression*, trans. Paul Kotman (Stanford: Stanford University Press, 2005), p. 9.

> Philosophers themselves—servants of the universal—are the ones who teach us that the knowledge of Man requires that the particularity of each one, the uniqueness of human existence, be unknowable. Knowledge of the universal, which excludes embodied uniqueness from its epistemology, attains its maximum perfection by presupposing the absence of such a uniqueness. *What* Man is can be known and defined, as Aristotle assures us; *who* Socrates is, instead, eludes the parameters of knowledge as science, it eludes the truth of the *episteme*.³

This thought connects with four thematic issues that inform the project of this book—epistemic vulnerability, incredulity, ignorance, and trust—issues that animate a commitment to crafting principles for constructing ways of knowing/being that foster the democratic, respectful cohabitation that I begin to address in *Epistemic Responsibility* and in *Ecological Thinking*.⁴ Here, I follow that thread to think about how these thematic modalities—vulnerability, incredulity, ignorance, and trust—influence and shape people's diverse ways of knowing and responding to diversity and "difference"; how they contribute to judgments about whose testimony merits a hearing, whose knowing merits attention for its innovative effects; whose knowing is thwarted in structures of incomprehension and intransigence. The "who" could be a single knower or a group, silenced, marginalized, enabled, validated, or celebrated. The putative knowledge could be an isolated empirical claim; it might be a complex theory or a hypothesis acknowledged, open to deliberation or contestation. These are just some of the possibilities.

Advocating an epistemological turn toward particularity evokes certain caveats. Its purpose is not to reclaim the individualism against which feminist, postcolonial, antiracist moral-political-epistemological critique has, appropriately, been directed. Nor is it to recenter simple empirical knowledge claims about cups on tables or cats on mats. Particularity, singularity and uniqueness are conceptually and ontologically distinct from individualism in their philosophical genealogy and in their nuances, significances, and effects. At its simplest, pursuing Cavarero's

---

3. Adriana Cavarero, *Relating Narratives: Storytelling and Selfhood* (London: Routledge, 2000), p. 9.

4. Lorraine Code, *Epistemic Responsibility* (Hanover, NH: University Press of New England, 1987); and *Ecological Thinking: The Politics of Epistemic Location* (New York: Oxford University Press, 2006).

thought asks us to revisit those old reminders that a language of pure particulars could be neither spoken nor understood. At the same time (to cite just two examples), a focus on particularity seems to ignore such apt warnings as Wendy Brown's caution against an "excessive specificity . . . [that] sacrifices the imaginative reach of theory" in what amounts, she believes, to a move toward positivism; and Robyn Weigman's wariness of a "particularist reduction whereby the . . . distillation of bodies from knowledge yields an understanding of identity studies as the sole institutional domain within which the complexity of power cannot possibly be thought."[5] From a different direction, it seems to advocate a return to the concrete, to that preoccupation with quotidian detail and care for the particular that, in Western-Northern societies, has long been women's lot, freeing men to occupy themselves with matters of universal import. Singly enacted, none of these consequences would be epistemologically or politically effective: neither generality nor particularity alone can motivate viable epistemic projects. In short, then, a turn toward particularity would have to navigate a difficult passage between the stark invisibility/inaudibility often generated by the *logos* as Cavarero construes it and the scattered dissolutions that moving too close to pure particularity could entail. These are among the questions that motivate my thinking in this book.

Now, particularity has not been entirely ignored in Anglo-American ethics, politics, and epistemology. Even orthodox, prefeminist *moral* judgments commonly come down to evaluating particular, specific actions with/to/on particular people, even though the practice itself tends to favor replicability. For feminist and other postcolonial moral theorists, engagement with the specificities of agency, detail, and situation is of vital significance in drawing attention to the omissions endemic to universalist analyses; nor *mutatis mutandis* will an insistence on paying attention to particularity, with the omissions it frequently exposes, be new to feminist and other postpositivist/postcolonial epistemologists. Objectivist practices in both domains, albeit differently, tend to gloss over gender-race-class specificities in ways that truncate their pertinence for nonwhite, nonmale, non"mainstream" knowers and activists.

Whereas orthodox twentieth- and early twenty-first-century Anglo-American epistemology—and, if to a lesser extent, ethics—has sought

---

5. Wendy Brown, "The Impossibility of Women's Studies," and Robyn Weigman "Feminism, Institutionalism, and the Idiom of Failure," both in Joan Wallach Scott, ed., *Women's Studies on the Edge* (Durham, NC: Duke University Press, 2008), pp. 31, 60.

to suppress particularity for its tendency to obstruct clear paths to the universal, feminist and other critical epistemologists and moral-political theorists have commonly started analysis from the particularities—the specificities—of women's lives and the lives of people otherwise Othered in the hegemonic discourses and practices of the mainstream. They have charted the epistemological, ethical, ontological significance of knowers' situatedness in particular—if multiple—relations to the circumstances, events, and people they seek to know, and in their relations to other knowers, singly and/or collectively. In its hospitality to the once-outrageous question "Whose knowledge are we talking about?" feminist epistemology has prepared the ground for addressing many of the issues Cavarero raises. But there are differences worth taking seriously into account.

The epistemology that has tended, if silently, to inform traditional Anglo-American moral theory has indeed sought to subsume the particular under general, universal precepts, if often appropriately so in the interests of impartiality and fairness. But the problem remains that the particular, the unique, the "case" that does not fit, tends often to disappear in such judgments for want, one might say, of taxonomic resources for addressing it. This disappearance is multiply troubling and especially so, as I will suggest in chapter 3, in its implications for questions of relationality and for the politics and ethics of testimony. Similarly, although there has been extensive work on moral particularism in Anglo-American philosophy, engagements with relationality, on which recognitions of particularity in some sense rely, have been less apparent in the tacit individualism that commonly informs such analyses. Complex ontological assumptions about multiplicity as somehow representing the specificities of diversity, with no need to take them separately into account, tend to be held apart from practices of analysis and justification. In consequence, the exceptions that traditionally "prove the rule" frequently disappear in the interests of the putative integrity of analysis.

One now-classic example illustrates these concerns. In his essay "Moral Perception and Particularity," Lawrence Blum, as his title suggests, develops an analysis sensitive to the particularities—the specificities—of moral situations, showing how subjective disparities in perceptions of salience feed into diverse moral judgments of situational specificity.[6] Yet in the ensuing contrast, Blum's analysis highlights a facet of Cavarero's

---

6. Lawrence Blum, "Moral Perception and Particularity" *Ethics* 101, no. 4 (July, 1991), pp. 701–25.

approach in which she parts company with, and speaks a different language from, many such occasions where particularity arises in Anglo-American philosophy, both feminist and otherwise Other. While Blum's fine analyses of moral perception, in this early work, follow a fairly typical third-personal pattern, showing how X perceives Y in certain particular (z) circumstances, Cavarero's recommendations are primarily addressive, relational, expressed in a language that recalls aspects of Annette Baier's and my own early arguments in favor of a language of *second-persons*: a language of "we," of "you and I," as contrasted with formal, impersonal locutions. Such a discursive shift is especially evident in the place Cavarero grants to "relating narratives": narratives that establish *relations* between and among people and that *relate* particular circumstances, feelings, and responses; narratives that convey the "who," which, in her view, tends to disappear into the "what" of third-personal tellings. The difference is not a minor one, yet the relationality Cavarero elaborates is no mere dyadic replica of the univocal, monologic singularity of impersonal, abstract individualism. It is from a reciprocal relationality such as she—following Hannah Arendt—advocates that social and political community and affiliation can be generated and sustained, between and among selves who bear only the most distant resemblance to the unified, frozen self of modernity. Somewhat enigmatically, and perhaps speculatively, contentiously, Cavarero suggests: "[W]omen are usually the ones who tell life-stories . . . like Penelope, they have since ancient times, woven plots with the thread of storytelling . . . Whether ancient or modern, their art aspires to a wise repudiation of the abstract universal, and follows an everyday practice where the tale is existence, relation and attention."[7] How such a proposal could connect with the materiality and concrete urgency of oppression and marginality in twenty-first-century Western societies is a question such seemingly peaceable narration may be challenged to address, but Cavarero's emphasis on mutual exposure, communication, and responsibility opens spaces for engaging with it.

In fact, Cavarero's thinking about relationality develops, in part, out of her reading of the place *natality* occupies in Arendt's thought, specifically in its attentiveness to the exposure of the newly born, an exposure Cavarero sees as ongoing, constitutive of human lives from birth to death. It is an exposure many philosophers of modernity, in their

---

7. Cavarero, *Relating Narratives*, p. 54.

press toward achieving autonomy and the self-protecting "buffered self" of modernity (to borrow Genevieve Lloyd's apt phrase), have sought to escape and indeed to deny.[8] This "self" fears the mutual exposure and openness of its "thrown-ness" into the world. Yet for Cavarero, "we" are fundamentally and ongoingly exposed to one another, and that exposure, either in attempts to paper it over and defend against it or in philosophies and practices that endeavor to know, understand, protect, and honor it, is instrumental in shaping human personal, social, and political interactions, all the way down. Thus, she suggests, still following Arendt: "Natality is the fundamental condition of every living-together and thus of every politics; Mortality is the fundamental condition of thought, in so far as thought refers itself to something that is as it is and is for itself. . . . [B]irth, action and narration become the scenes of an identity that always postulates the presence of an *other*."[9] As I read it, introducing mortality into this picture is neither a negative nor a nihilistic choice: it attests, rather, to the real-world temporality of all human action, thereby adding an element of significance and perhaps urgency to epistemic interactions in their inevitable "situatedness." Critiques of the widespread implications of a Heideggerian focus on being-toward-death, in the writings of such feminists as Mary O'Brien, Grace Janzen, and Patricia Johnson, are just some of the voices that attest to how *assuming* (in Beauvoir's sense) and working with natality can open the way to a listening such as Cavarero advocates, out of which sensitive reciprocal *intelligibility* across diversity and resistance might be initiated.

My interest in engaging with these ways of thinking is prompted, if somewhat incongruously, by work in Anglo-American social epistemology,[10] which I engage in order to think away from places where

---

8. Referring to Charles Taylor's talk of "the 'buffered' disenchanted self" in the modern world, Lloyd argues that the "loss of providence exposes that buffered modern self to new vulnerabilities," in Genevieve Lloyd, *Providence Lost* (Cambridge, MA: Harvard University Press, 2008), p. 322.

9. Cavarero, *Relating Narratives*, p. 28.

10. C. A. J. Coady, *Testimony: A Philosophical Study* (Oxford: Oxford University Press, 1992); Michael Welbourne, *Knowledge* (Montreal: McGill-Queen's University Press, 2001), and *The Community of Knowledge* (Aberdeen: Aberdeen University Press, 1986); Jennifer Lackey and Ernest Sosa, eds., *The Epistemology of Testimony* (Oxford: Clarendon, 2006). See Lorraine Code, "Testimony, Advocacy, Ignorance: Thinking Ecologically about Social Knowledge," in Adrian Haddock, Alan Millar, and Duncan Pritchard, eds., *Social Epistemology* (Oxford: Oxford University Press, 2010), pp. 29–50.

philosophers treat testimony as consisting in information-conveying one-liners reporting everyday "facts" (often to a single interlocutor) and to think toward considering other modalities of testimony where it functions as telling, as exposure, as involving trust and vulnerability, as communicatively interactive. Such elaborated testimony is enacted in talking back and forth, creating relations between teller(s) and hearer(s) that in their political implications, be they benign or malign, exceed the one-liners uttered on a (presumptively) level playing field, on which Anglo-American testimony literature often relies.

A point of entry into reading this work with and against a background of Cavarero's thinking is to move away from a spectator mode of knowing, toward an engaged, addressive mode such as is adumbrated in Miranda Fricker's early work on epistemic injustice, and in her indebtedness to Edward Craig. In my view, it is partly owing to the centrality it accords to testimony and to simple and complex knowledge-conveying exchanges between and among people in the real world that much twenty-first-century Anglo-American social epistemology claims its title, a feature evident in the attention many social epistemologists give to extended examples of explicitly situated and populated epistemic negotiations, and—for my purposes here—in the subtle yet far-reaching effects of linguistic shifts from impersonal, third-person propositional claims that "S knows that p," to the language of speakers and hearers.[11]

Where reliance on the formal S-knows-that-p rubric enabled twentieth-century Anglo-American epistemologists to imagine they could transcend the vicissitudes of the world by establishing universal, necessary and sufficient conditions for the existence/achievement of "knowledge in general," Craig, as his title reference to the "state of nature" signals, returns to the world as the multifaceted place where knowledge is made, conveyed, deliberated, and adjudicated. A distinction he draws early in the book is germane to the epistemic significance of this shift to an addressive mode. Craig observes, "There are informants, and there are sources of information . . . Roughly, the distinction is between a person's telling

---

11. Edward Craig, *Knowledge and the State of Nature: An Essay in Conceptual Synthesis* (Oxford: Clarendon, 1990); Adriana Cavarero, *For More than One Voice: Toward a Philosophy of Vocal Expression*, trans. Paul A. Kottman (Stanford: Stanford University Press, 2005); Miranda Fricker, *Epistemic Injustice: Power and the Ethics of Knowing* (Oxford: Oxford University Press, 2007); José Medina, *The Epistemology of Resistance: Gender and Racial Oppression, Epistemic Injustice, and Resistant Imaginations* (Oxford: Oxford University Press, 2012).

me something and my being able to tell something from observation of him."[12] Craig focuses throughout the book on practices of hearing and speaking as these are enacted in people's engaging with one another as informants and interlocutors (albeit, mostly one-on-one). By contrast, treating people as sources of information involves a level of objectification not unlike what is involved in knowing the age of a tree by counting its rings (his example). Yet it is precisely in the attention he accords to the informant that the innovative aspect of Craig's approach is most apparent. This is no mere spectator theory, but a communicative one, where people acquire knowledge from and with one another "as subjects with a common purpose, rather than as objects from which services, in this case true belief, can be extracted."[13] That said, most of Craig's discussion is of third-person and first-person knowledge. There is little talk of second persons beyond this reference to a common purpose.[14] Hence a flavor of replicability remains, references to informants notwithstanding: a sense that no epistemic specificities of claimant, situation, or circumstance are germane to the cognitive exchange. In this aspect, his analyses are situated in an epistemic space quite different from Cavarero's. More significantly, his discussion of informants makes no mention of the *politics* of testimony, of asymmetrical social-epistemic positionings of speaker and hearer, or of the multiple ways in which treating people as sources of information could play into their vulnerability, their epistemic erasure, as readily as these might enhance their claims to know.

Recalling Cavarero's remark about philosophies that value dialogue ignoring the relationality of reciprocal communication, we see something similar reflected in Craig's failure to consider the possibility that dealing with people as informants will be as smooth and straightforward as he suggests *only if* speaker(s) and listener(s) are situated on a level playing field, with equal expectations of claiming a place to speak and of being heard and believed. Beyond simple exchanges of neutral information—if indeed in those—such ease of communication can rarely be assumed. Working to achieve some measure of it is central to responsible inquiry. To dispel a misconception I may have generated about Cavarero's work, from my selective reading, it might appear that she *also* concentrates on

---

12. Craig, *Knowledge and the State of Nature*, p. 35.
13. Craig, *Knowledge and the State of Nature*, p. 36.
14. See e.g. Craig, *Knowledge and the State of Nature*, p. 68.

narrators thinking and talking in comfortable exchanges and in polite, stable societies, to the neglect of conflict, marginalization, and oppression. To counter that impression, consider the quotation from *For More than One Voice*, with which I began:

> The revaluation of the vocalic that I am here proposing, although it has good reasons for opposing itself to the history of political logocentrism, does not aspire to a definitive liberation from politics, but rather posits a different way of thinking the relation between politics and speech. As Arendt invites us to consider, the question lies precisely in thinking the elementary criterion of a politics that valorizes the relationality of the unique beings that manifest themselves actively through speech, leaving aside the imperialism of the Said . . . [T]he protagonist of this politics is a speaker who, leaving aside his or her belonging to this or that identity group, this or that language, communicates him- or herself first of all as voice.[15]

An implicit dissatisfaction with individualist, monological, dislocated one-liner epistemology is apparent in these observations, and rightly so.

That said, it must be acknowledged that a large part of everyday testimonial exchange in the Western-Northern world does in fact deal with one-liners in which items of information are passed from speaker to hearer (or from writer to reader). This practice is worthy of discussion on its own terms; but the mistake is to regard these events as paradigmatic for thinking about "testimony in general." On occasion, it may well merit that status. Nevertheless, such accounts are rarely sufficiently rich or capacious to allow entry to testimonial evidence that, at least in its first uttering, is radically particular and often inchoate, tentative, and exposed in the vulnerability from which it comes and which, in risking incredulity, it could exacerbate rather than alleviate. The tellings it utters may find no place in a received social imaginary where hermeneutical resources are lacking to allow them an adequate hearing or even to initiate the kinds of negotiation that might ultimately produce an effective hearing.

Closer in some respects to Cavarero's position is Fricker's *Epistemic Injustice*, whose subtitle *Power and the Ethics of Knowing* suggests a departure

---

15. Cavarero, *For More Than One Voice*, pp. 200, 209.

from sedimented presumptions of epistemic neutrality. Fricker's attention to particularity is apparent in the analyses of literary works she invokes to illustrate what is at stake in the examples of hermeneutic and testimonial injustice that are her focus. There, she begins to examine the potential of delineated, narrated examples to yield descriptions more akin to (some) phenomenological analyses in their capacity to disclose the implications and effects of epistemic injustice. Making hermeneutic injustice intelligible, exposing the lacunae in the hermeneutical repertoire on which some of this discussion relies, cannot be achieved in simple attenuated one-liners. Nonetheless, and despite Fricker's claims to the contrary, the *hermeneutic* aspects of such a new focus are the more significant of its conceptual figures: testimonial injustice is itself enabled or thwarted by hospitable/inhospitable hermeneutic environments.

While these matters are, in many ways, as old as Western philosophy itself, they take on a particular urgency in this highly bureaucratized-digitalized-mechanized twenty-first-century world where so many of the categories into which people must fit themselves, and into which the unfamiliar, the strange are required to fit, are too crude, too limited and limiting to enable recognition of those thus categorized as more than a "what." In the terseness of their articulation, they truncate possibilities of knowing well enough to inform "ways of going on," of understanding the complexities and demands of specific situations, actions, knowers. These limitations attach in part to the crudity of stereotypes and also—to give just one example—to cases such as I cite in *Ecological Thinking* with reference to biologist Karen Messing's research on women's workplace academic granting agencies, where the nonquantifiable character of some workers' stories of their symptoms casts them as too particular, too variable in their detail to yield appropriately credible results. They slip through the cracks of apt acknowledgment, and hence of certain compensation schemes. Such situations need to be addressed and remedied.

Nonetheless, it would be overly simple to propose, at this point, that narratives such as Cavarero invokes are the solution to the problems that puzzle us. Reliance on testimony has long occupied a low rung on the ladder of practices designed to produce knowledge good of its kind, whether justifiedly or not. Whereas a claim that the cup is on the table is easily verifiable in most "normal" circumstances in the affluent Western-Northern world, more complex testimonial claims are rarely verifiable in single "look-and-see" situations. Determining the adequacy of a testimonial claim—perhaps of a mininarrative—can take time and require ongoing (if intermittent) investigation. Some of these claims, such as affirming the effectiveness of a drug, a pesticide, a system of voting, a gender-sensitive politics, require ongoing and wide-ranging inquiry, even as people act upon "state-of-the-art" findings. There are situations where an initial openness—a measure of credulity—can be the only way to "go on," paradoxically leaving a "final" conclusion open to continuing investigation. A sense of the openness of much everyday knowledge, together (paradoxically) with a need for persistent, reasonably open-ended inquiry, is a contradiction that accompanies much quotidian epistemic judgment, now. In short, no single narrative can tell "the whole story," even if the very idea of a whole story were intelligible. Hence, none of the examples and caveats cited here point toward simple either/or choices: thinking well about them produces more questions than answers. For a start, with regard to knowing other people, it is not always or even usually certain that "the Other," "the oppressed" *wants* to be known in her, his, or their particularity. People are private, often closed, and with reasons that have to be respected, despite the apparent implausibility of the "buffered self."

This last thought recalls a remark from Cynthia Cockburn in a different situation, about interviewing conflict survivors, where she asks readers to recognize that those conducting the research have to be mindful that "the researched take greater risks and lack the mobility, resources and choices available to the researcher" who, if she would know them responsibly must respect their not-telling as fully as she respects their telling.[16] Too briefly, then, her warning recalls Alison Bailey's early appeal to liminality in her reading of Maria Lugones's "curdled logic"[17] and to the need to respect "our" opacity one to another thus enjoined. Feminist epistemologists are commonly positioned to acknowledge such opacity, with the consequence that their/our epistemic practices may fare better than standard mainstream Anglo-American epistemology can, which lacks the conceptual tools required to acknowledge and capture this phenomenon. Feminist epistemology is committed to developing resources with capacities for acknowledging such opacity, whereas in the formulaic knowledge claims that are the stuff of which Anglo-American

---

16. Cynthia Cockburn, *The Space between Us: Negotiating Gender and Identities in Conflict* (London: Zed Books, 1998), p. 3.

17. Alison Bailey, "Strategic Ignorance," in Shannon Sullivan and Nancy Tuana, eds., *Race and Epistemologies of Ignorance* (Albany: SUNY Press 2001), pp. 153–72.

mainstream epistemology is commonly made, it is less plausible to articulate contentions of this nature. But even opacity within exposure cannot readily be conveyed. Understanding, engaging with such opacity requires commitments to listening and negotiating across differences, even if negotiation results in a deeper commitment to preserving silence, to "holding open a space for not knowing," a modality that does not reduce to ignorance, but to reserving judgement, waiting to see.[18] Testimony, whether thinly or thickly articulated (i.e., in one-liners or in extended tellings) is often so specific to a situation, to a "who" (singular or plural) in that situation that if it is to contribute without doing epistemic violence to the intelligibility of the experiences told, it needs, perhaps impossibly, to be heard and accorded an initial presumption of credibility in its uniqueness, even as hearers and speakers strive to achieve a meeting place on uncharted epistemic territory. Such problems are further exacerbated by a realization integral to feminist, antiracist, phenomenological, and postmodern thinking that particulars, whether material or human, are neither unmediated in their particularity nor unmixed in a liberal sense of presenting unified transparent wholeness. Nor are they stable, fixed, or once-and-for-all. So, knowing particulars is at once urgent and, in any definitive sense, impossible. Knowledge is always unfinished and incomplete, precisely because of the open-endedness of experiences and meanings.

With these observations about particularity and wholeness, I have not begun to approach a conclusion, but I bring them to an (interim) ending with two disconnected observations that point toward one direction this line of thinking could take. First, in a sense, these thoughts revisit the old problem of the one and the many, of mediaeval debates about "thisness," ancient questions about universality and particularity. To reiterate my reminder about a language of pure particulars, it is clear in obvious ways that abstraction is usually essential for communication to take place. The problem is to determine when abstraction abstracts too much, moves too far away from the concrete, from particular harms, neglects, and urgencies, and in so doing sanitizes them to the point of insulating them against the recognition, the acknowledgment, without which they cannot be brought to the level of intelligibility requisite for addressing them adequately. Second, here again, the implausibility of

---

18. Bailey, "Strategic Ignorance," p. 160.

the solitary and self-contained knower is apparent. Acknowledging the epistemic centrality of testimony, in its multiple modalities, reinforces this contention. Yet my initial claim, following Cavarero, is also more enigmatic than I first suggest, because it is after all not true, in one sense, that philosophy does not deal with particulars. Philosophy itself is a social institution that comes to all it does, all it knows, all it classifies and discusses, with applicable—or not—categories and classifications in place, if engagement is to be possible at all. The conundrum Cavarero poses is about how to retain and honor particularity while enlisting resources ready at hand, if perhaps not yet readily available, to work toward its intelligibility, not as what but as a *Who*.

The Myth of "The Individual"

Who *is* this individual? this autonomous moral-epistemic agent? this exemplary/typical knowing, acting, suffering or thriving human being? Such questions in diverse modalities, originating in multiple circumstances and interests, have long been virtually invisible in Anglo-American philosophy in general, in epistemology in particular, and even in social/political/moral philosophy. He—for he is tacitly assumed to be a featureless male place-holder—tends not to appear as "*who* he is" for it has seemed reasonable, in the interests of maintaining detached objectivity, to work with a presupposed "generic" subject as moral agent, as knower, as sufferer, as living or dying. For Anglo-American epistemologists and moral theorists, this subject has commonly been an undifferentiated, unidentified placeholder across established analyses of knowledge, subjectivity, morality, belief, and agency. Specificities, particularities have tended to count merely as impediments to analysis good of its kind, as incidental, as vehicles for carrying evaluations and putative understandings forward; as populating general theories and generic subjectivities. Concentration on the specificities of *their* detail is commonly avoided in the interests of universality, of neutrality. None of these observations are new, but they take on a renewed urgency in analyses such as these, which invoke assumptions about a putatively generic humanity and agency into which such specificities, such individual details tend to disappear, often precisely on account of their particularity. It is an odd conundrum.

Feminist and other nonmainstream philosophers have engaged critically and constructively with the complexities of an over-inflated esteem for autonomy and with some of the problems inherent in a "received" conception of "the individual." So too, if obliquely, have Jonathan Beevor and Nicholae Morar, notably in their provocative essay "Bioethics and the Challenge of the Ecological Individual."[19] These authors engage carefully with philosophical endeavors in the Anglo-American tradition, to arrive at persuasive conclusions about the truncating effects of prefeminist, prepostcolonial "individualism" as it pertains to patients in biomedicine and to wider populations. In many ways, the very ideas of an "individual" and *a fortiori* of an "autonomous individual" are either question begging or oxymoronic in the unarticulated ontological assumptions about human embodiment, biology, agency, situation, and maturation they silently carry. Quite simply, and now by no means a new thought, we human creatures could neither be *tout court*, nor even strive to be autonomous "individuals" in any but highly attenuated and indeed fanciful imaginary ways. Such, in large part, is the message I convey in my 2006 book *Ecological Thinking: The Politics of Epistemic Location*, on which these authors draw extensively and generously in their text. Well-known studies of "wild children,"[20] together with analyses of child development and social deprivation such as I discuss in chapter four of *Ecological Thinking*, begin clearly to show how the putatively "autonomous individual" is more artefact than fact and to demonstrate that the circumstances of its making do not culminate "naturally" in achieved autonomous individuality.[21] We, whoever we are, are creatures of relationality, of multiplicity and interdependence—be it with/on other human beings, on other animate beings, on "our" surroundings, on the offerings/resources/limitations of the world as we find it, however depleted or plentiful they may be. Theoretical projects that suppress or step aside from acknowledgment, in Wittgenstein's sense, of such interconnectedness cannot, in my view, achieve adequacy. Hence, to emphasize the fundamental significance of this point, in *Ecological Thinking*, I strongly endorse Anne Seller's

---

19. Jonathan Beever and Nicholae Morar, "Bioethics and the Challenge of the Ecological Individual," In *Environmental Philosophy* 13, no. 2 (2016), pp. 215–38.

20. See Adriana Benzaquén, *Encounters with Wild Children: Temptation and Disappointment in the Study of Human Nature* (Montreal: McGill-Queen's University Press, 2006).

21. Lorraine Code, *Ecological Thinking: The Politics of Epistemic Location* (New York: Oxford University Press, 2006), pp. 129–61.

claim that "as an isolated individual, I often do not know what my experiences are"[22] and Ludwig Wittgenstein's "knowledge is in the end based on acknowledgement."[23]

Such claims were even more difficult to articulate with any expectation, indeed, of acknowledgment, prior to the birth of social epistemology in English-language philosophy in the late twentieth century: a development I again attribute, largely, and not arbitrarily, to Edward Craig. I read his analysis as innovative, especially for its introduction of the language of speakers and hearers as these contrast with isolated individual utterances into a void that are the stuff of which much presocial, prefeminist, predifference-cognizant Anglo-American epistemology is made. In my view, so-called "continental" engagements with knowledge matters have worked with more nuanced, more specificity- and situation-aware, more affect-friendly conceptions of human being than mainstream twentieth-century analytic philosophy has done. In these respects, some nonanalytic readings are germane to the way Beever and Morar have taken up the quest for a viable departure from an inflated veneration of autonomy such as has, for example, informed mainstream medical thinking and practice in the affluent West/North. Their readiness to entertain the suggestion "that we should not think anymore of ourselves (and of other organisms) as *individuals* . . . but as *communities* or *ecosystems*" is congenial to this way of thinking.[24]

But, in one significant respect, I part company with Beevor and Morar. As they rightly note, in *Ecological Thinking*, I represent my approach to ecology as both literal and metaphorical, and I concentrate more specifically on the metaphorical than they propose doing. I neither conceive of nor represent these two "ways" as antithetical one to the other. In both, it is apparent that knowers participate, albeit diversely, as shapers of a going episteme. My focus is on the metaphorical, as they observe, but it is not obvious that the two "ways" are mutually exclusive.[25] They can often, if not always, be reciprocally informative, although I am not sufficiently knowledgeable, scientifically, to engage as fully as the authors do, with the literal way. Here, a valuable resource

---

22. Code, *Ecological Thinking*, p. 174.
23. Ludwig Wittgenstein, *On Certainty*, #378.
24. Beevor and Morar, "Bioethics," p. 10.
25. Beevor and Morar, "Bioethics," p. 12n6.

is Theodore Brown's 2003 *Making Truth: Metaphor in Science*,[26] together with a (still germane) 2004 conversation in which I contend—as I would again—that "understanding how things get naturalized is a significant piece of responsible knowing. The hardest expectations to see are those that have been so thoroughly naturalized that they become part of the fabric of everything we think, and everything we think we know."[27] Clearly, the literal and the metaphorical implications of thinking/knowing ecologically can be so closely interwoven as to be unequivocally, reciprocally interconnected in shaping the cognitive ecosystems in which "we" know and think and live. Participation in such understandings, especially for lay readers and thinkers, can be enabled through apt, metaphorically shaped arguments, which need not, and frequently do not, lose sight of the literal.

Of a piece with the ontological assumptions these thoughts invoke is a conception of all-or-nothing, perfectly definitive knowing—absolute certainty, and nothing less—characteristic of a postpositivist epistemic imaginary that tacitly governs such thinking, in effect making it impossible *ever* to know well enough to give up on presumptions of entitlement. It exposes the power of cultivated, sustained ignorance to generate a complacency that "justifies" damaging practices, be they environmental or social or a mix thereof. While not all the examples analyzed in these essays are explicitly about ecologically pertinent matters, the structural implications of ignorance nourished and sustained translate well into analyses of climate change skepticism.

In a 2014 article, Naomi Scheman speaks to and from a compelling engagement with questions such as these. The essay makes a vital contribution to thinking across the putative divide separating the metaphorical and the literal, as Scheman invites her readers to consider the implications of maintaining that:

> [A]ll objects of knowledge (including, e.g., persons, plants, and cells) are constituted relationally and are knowable through the relationships in which they are embedded. Laboratory-based

---

26. Theodore L. Brown, *Making Truth: Metaphor in Science* (Urbana: University of Illinois Press, 2003).

27. The conversation is published as Kirsti Malterud, Lucy Candib, and Lorraine Code, "Responsible and responsive knowing in medical diagnosis," *NORA* 12:1 (2004), pp. 8–19; I quote from p. 14.

research, and research that takes laboratory science as an aspirational paradigm, abstract objects of knowledge from the relationships in which they are embedded and re-embed them in relationships constituted by the research setting itself. For research findings to be applicable outside the laboratory, "in the wild," knowledge is needed about the relationships from which the objects were abstracted, knowledge typically unavailable to the researchers unless they are respectfully—sustainably—engaged with the diverse communities from which their objects of knowledge have been abstracted and into which they will be reinserted.[28]

In short, Scheman is contending that objects of scientific knowledge can be at least partially constituted by their relationships in the research setting, be it "naturally," institutionally, or otherwise situated. The extent and influence of their "situatedness" will vary place to place, with the detail of place often being determined, and evolving, case by case. Even ecological research is, at times, conducted in a laboratory setting. Often the objects' entry/return into the "everyday world" will bear traces of those encounters, places, situations, whether negatively, neutrally, or positively. This claim, with which I concur, lends support to the contention that the literal and the metaphorical can be reciprocally constitutive. The extent to which such a view claims explanatory power will vary case to case, but it supports my earlier contention about the reciprocity of the literal and the metaphorical, despite my less than credible knowledge of the literal. The takeaway point for wide-ranging inquiry—even for the traditional "knowledge in general"—is that both knower and known are embedded in a literal and a metaphorical ecosystem whose effects inevitably, and by no means negatively, contribute to shaping both process and "product."

To conclude these thoughts on a promissory note, if not with a conclusive conclusion: again, there is a rich, productive, feminist-informed *ontological* resource for biomedical research and practice, and for ecological thinking generally, in Cavarero's *For More than One Voice*. She maintains that "we are beings who are, of necessity, *exposed* to one another in our

---

28. Naomi Scheman, "Empowering Canaries: Sustainability, Vulnerability, and the Ethics of Epistemology," *International Journal of Feminist Approaches to Bioethics* 7, no. 1 (Spring 2014), pp. 169–70.

vulnerability and singularity, and . . . our political situation consists in part in learning how best to handle—and to honor—this constant and necessary exposure."[29] I am proposing that such recognition is fundamental to thinking ecologically and to knowing responsibly in their interactive, literal, and metaphorical modalities.

Culpable Ignorance?

In a chapter titled "Epistemic Responsibility and Culpable Ignorance," José Medina asks: "When is partaking in a body of social ignorance a form of irresponsibility? . . . And is the failure in responsibility an ethical failure of the individual or a political failure of society?"[30] The questions are timely. Appeals to epistemic responsibility, which had been something of a sleeper since my book by that title was published in 1987, are enriching the conceptual repertoire of Anglo-American social epistemology, opening new rhetorical-discursive spaces.[31] Such appeals acquire an enhanced urgency in relation to climate change skepticism, with the doubts that feed it and are nurtured to preserve it. Participating in such skeptically generated social ignorance is indeed, and always, a form of irresponsibility, for the consequences of denial are multiple and dire, not just for "the environment" but for lives both human and other-than-human, throughout the world. Ignorance of such matters is culpable. It is at once an ethical and a political failure, with ethics and politics reinforcing one another. It is, primarily and overwhelmingly, an egregious failure of epistemic responsibility, with widely cultivated-manufactured ignorance and doubt sustaining the ethics and politics that require contestation. Yet the question *whose* irresponsibility is at issue and how it could/should be addressed is complex and fraught in a time of putatively conflicting information that few "ordinary people" are equipped to disentangle from the vested interests and unstable expertise that infuse it. Answering the "whose responsibility?" question definitively presupposes that ignorance is recognizable and that its "partakers" are positioned to acknowledge it as such. Such presuppositions are highly contestable.

---

29. Cavarero, *For More than One Voice*, p. 9.
30. Medina, *The Epistemology of Resistance*, p. 133.
31. Code, *Epistemic Responsibility*.

Consider just one among many examples that both complicate and explicate these questions. The 2012 film *Chasing Ice* is billed as the story of one man's mission to change the tide of history by "gathering undeniable evidence of climate change."[32] The driving assumption is that when people *see* the evidence, their skepticism about "global warming" will be silenced: they will *know*. Acclaimed *National Geographic* photographer James Balog was a climate change skeptic until, in his Extreme Ice Survey, he uncovered compelling visual/corporeal evidence of how the planet is changing. His videos compress years into seconds as they capture ancient mountains of ice in motion, disappearing at a breathtaking rate. Informed and mobilized by what they will have learned, he hopes people will understand the urgency of promoting climate change awareness and activism, that the films will contribute to unsettling social ignorance in its most blatant aspects, and stimulate committed engagement. While I hope he is right, my response is a mixture of concern, dismay, and astonishment at the film's sheer daring and beauty. The concern is prompted by the "undeniable evidence" and also—if perversely—by questions about how seriously it will count as "undeniable" in a public forum, where ardent naysayers insistently, relentlessly succeed in promulgating doubt and attracting massive financial support for their endeavors. In an era when cinematic special effects seem to know no boundaries, it is easy to imagine diehard deniers denying even this demonstration that change is dire, rapid, and frightening, that it calls for urgent epistemic, social, political, and ethical responses. But—if incongruously—even after viewing *Chasing Ice*, where the cinematic evidence is astonishing, it is hard to resist thinking—fearing—that it would not be a significant stretch for naysayers and deniers to read the evidence the film presents, at such high cost and dedication, as just another display of amazing "special effects." Balog, an accomplished and eloquent environmentalist, might well be horrified at such a thought; but it is not so fanciful.

How do these thoughts bear on my unequivocal response to Medina's questions about the irresponsibility or otherwise of partaking in social ignorance, about when or whether social ignorance counts as a form of irresponsibility? At the very least, they complicate the situation beyond the complexity that attaches to straightforward epistemological appeals to truth or falsity. This epistemic confusion is endemic to politics-of-

---

32. *Chasing Ice*, directed by Jeff Orlowski, written by Mark Monroe (2012).

knowledge issues that are multiply tangled in their implications for projects of developing reasoned, informed ways of countering climate change skepticism and addressing ecological ignorance. It almost goes without saying that countless members of affluent societies have much at stake in preserving the status quo with the alleged ignorance that sustains it, despite its ruinous consequences. Perhaps with such thoughts in mind, Val Plumwood argues: "We need above all an *ethical* science" because "sado-dispassionate science has used the ideology of disengagement to wall itself off from ethics."[33] Feminist and other social epistemologists have been engaged in dismantling this "wall," and matters of epistemic responsibility are central to such deconstruction and reconstruction projects. Medina's question, I propose, urges us also to attend, in these projects, to such nonstandard, nonpropositional sources of evidence as *Chasing Ice* or to Judith Butler's reading of the Rodney King case and Andrew Reszitnyk's photographs of the consequences of an oil-rig explosion, to expand challenges to sedimented ignorance beyond conflicting verbal assertions, to showing as well as telling.

Yet looking to Balog's film for undeniable evidence—for "knowledge" that can dispel ignorance—raises questions that expose the contestability of the old empiricist "seeing-is-believing" adage. Even with perceptual "givens" in the world of Heidegger's *allgemeine Alltäglichkeit* (average "everydayness"),[34] we do not always know what we are seeing: interpretation, reviewing, and deliberative discussion are vital to understanding such empirical-observational reports, which rarely wear their meanings on their sleeves. Thoughts such as these pertain analogously to other modalities of visual imagery—paintings, photographs, films—when they are adduced as sources of knowledge. Often, "we" literally cannot see what is before our eyes, even in circumstances that rely on the alleged authority of direct perception; nor can visual perception be presumed to be as univocal or as "neutral" as postpositivist empiricists suggest. Analogously challenging, then, in situations that pit naysayers and environmentalists against one another, is Judith Butler's reading of the Rodney King case, where she writes of "reproducing the video within a racially saturated field of visibility," a situation that, in its susceptibility to multiple readings, shows affinities with reproducing *Chasing Ice*

---

33. Val Plumwood, *Environmental Culture: The Ecological Crisis of Reason* (London: Routledge, 2001), p. 53.
34. Martin Heidegger, *Being and Time* (New York: Harper & Row, 1962), p, 421.

within a climate change-skepticism saturated field of inquiry.³⁵ As Butler notes, "there is no simple recourse to the visible, to visual evidence . . . [I]t still and always calls to be read."³⁶ In short, it may on occasion be plausible to assume that seeing is believing, but the product of reading such evidence cannot, without further ado, be equated with "justified true belief," which becomes knowledge and thus dispels uncertainty.

Analogous issues attach to readings of photographic evidence of the 2010 Deepwater Horizon oil-rig explosion in the Gulf of Mexico, where, Andrew Reszitnyk reminds us, "photographs are not transparent representations, but active interpretations of it."³⁷ Charley Reidel's photographs—the focus of Reszitnyk's discussion—show animals and birds coated with, and likely dying from, the oil on the water. There are disanalogies with *Chasing Ice* in the contrasts between animate and inanimate subject matters, but there are continuities, especially with regard to viewers-as-putative-knowers' presumptive responsibilities to investigate and deliberate cautiously before claiming to know/believe from these seeings and to evaluate Reszitnyk's admittedly contestable admonition: "[W]e are, in a way, obligated to respond to and take responsibility even for disasters that seem to have nothing to do with human activity."³⁸ His point is moot, if not implausible. It is multiply convoluted with respect to *Chasing Ice,* where it is less easy for the uninitiated to attribute causal factors to human activities than in the *Deepwater Horizon* situation, again, with respect to how such responsibilities should be enacted. But the claim is well taken, if only because of the complexity—even with *Deepwater*—of drawing precise borders between "natural" disasters for which, allegedly, no one is responsible, and disasters where human agency is incontrovertibly at work. Vital for fulfilling the responsibilities Medina invokes is a readiness to take seriously the diverse arguments and demonstrations advanced, to the effect that activity far away—geographically, demographically, temporally—can and often does significantly impact ice and other hitherto putatively inviolable natural phenomena. The point is not to shift from "seeing is believing" to "hearing expert

---

35. Judith Butler, *Bodies That Matter: On the Discursive Limits of Sex* (New York: Routledge, 1993), p. 15.

36. Butler, *Bodies*, p. 17.

37. Andrew Reszitnyk, "Eyes through Oil," unpubd. PhD Diss, McMaster University (2012), p. 145.

38. Reszitnyk, "Eyes through Oil," p. 150.

judgements is believing." But such judgments, whose genealogy must continually be evaluated both laterally and vertically (i.e., transversely across current issues, and historically) invite us—urge us—to stop and investigate responsibly even about *prima facie* implausible contentions. Plumwood in this regard has high hopes for "ecological education and institutional change, to develop in the culture the right sorts of prudential and care-based, non-ephemeral reasons for considering nature's interests."[39] Although the criteria for recognizing what counts as "right" are elusive, her claim anticipates my second example.

In an "open society" where an assumption prevails that accurate knowledge/information is universally available, and where its "consumers" tend to believe that everyone can and should take due cognizance, engaging Medina's questions about social ignorance unequivocally, in the affirmative, seems to require no argument. But responsibilities are interwoven and reach far beyond the readership of philosophy books and scientific articles or the credulity of filmgoers. In the affluent Western-Northern world now, many of these responsibilities are also *pedagogical*: educators, investigative journalists, and public intellectuals (among others) have a presumptive duty to know, address, communicate, and debate these issues in their complexity; and responsible citizens have some obligation to learn how to evaluate them, negatively or positively. Yet assuming such responsibilities is, again, a fraught, often frustrating, task; and questions about where to confer trust are not easily addressed. For Plumwood, "[t]hrough local education, activists can stress the importance and value of nature in practical daily life" can demonstrate "the imprudence of anthropocentrism . . . by showing the extent of uncertainty and the limits of our knowledge" as a counter to "arrogance wrapped in the garments of science."[40] But, even if they/we recognize these limits, how can activists know we/they are right?

Again, the evidence is anecdotal. I am thinking of a colleague who, to my asking what he expected his students *to know* about the urgency of climate-ecological issues, responded: "Virtually nothing: their main source of information is Fox News." Just as urgently, I am thinking of textbooks approved for use in schools as presumptively neutral sources of "objective" knowledge-as-information, a dominant assumption in "open

---

39. Plumwood, *Environmental Culture*, p. 142.
40. Plumwood, *Environmental Culture*, pp. 112–13.

societies." It conflicts, however, with a report of Texas state legislators' part in determining how science should be taught, where the politics of knowledge that informs such decisions leaves teachers and would-be knowers at a loss as to how responsible epistemic practice is to be achieved. For example, and again anecdotally, Gail Collins reports:

> Approval of environmental science books was once held up over board concern that they were teaching children to be more loyal to their planet than their country. As the board became a national story and a national embarrassment, the state legislature attempted to put a lid on the chaos in 1995 by restricting the board's oversight to "factual errors." *This made surprisingly little impact when you had a group of deciders who believed that the theory of evolution, global warming, and separation of church and state are all basically errors of fact.*[41]

Pedagogical responsibilities are multiply challenging in such situations where the tacit agenda is apparently one of producing and sustaining some version of what Shannon Sullivan calls "ignorance/knowledge."[42] How can teachers, who may themselves have been taught in systems of climate change denial, who do not think to, or may be afraid to, or believe themselves insufficiently expert to question the presumptive "authority" of texts and the officials who endorse them, hope to counter textbook accounts in what is evidently a closed, tightly monitored power/knowledge system, masking the agendas that animate it? Nor is it insignificant that some teachers, in situations where fragile employment conditions prevail, would be wary of rocking the boat or that at primary and secondary levels, at least in many parts of the affluent world, teachers are overwhelmingly female, hence multiply vulnerable: to the force of sex/gender stereotypes, and to a level of financial instability that is often race-, class-, and gender-related in interwoven ways that threaten their professional status, authority, and security. How might they respond to Medina's questions? With variations from situation to

---

41. Gail Collins, *As Texas Goes . . . : How the Lone Star State Hijacked the American Agenda* (New York: Norton, 2012).
42. Shannon Sullivan, "White Ignorance and Colonial Oppression: Or, Why I Know So Little about Puerto Rico," in Shannon Sullivan and Nancy Tuana, eds., *Race and Epistemologies of Ignorance* (Albany: SUNY Press, 2007), pp. 153–72.

situation, tolerance of nonconformist action can be scarce. Even in unionized situations where jobs may be securely protected, prospects for advancement and for various forms of professional engagement outside the school itself are—often tacitly—limited. With local variations, whistle blowing and multiple other forms of protest will be an option only for the most secure[43] in societies where endless resources must be mustered to substantiate charges, and where whistle-blowers (female and male) are more often judged subversive, disloyal to their employers and colleagues, more likely to meet with destructive opprobrium and/or to lose their jobs than to garner respect as people of conscience exercising their responsibilities to know and to circulate their knowledge, against the odds, these questions are by no means merely rhetorical.

Feminist, antiracist, and other successor epistemologists have created spaces for thinking about matters such as these, if perhaps not in ways that would convince the deniers, given the opaque structures of vested interest that fuel and reinforce their cultivated, well-rehearsed skepticism. Moreover, one-off individual denials, as Michel Foucault shows, are too-readily dismissed as anomalous, evidence of craziness, in a power/knowledge system where the word of authority is not easily gainsaid and contesting it is less often heard as indicative of "knowing better," more often as indicative of being dangerously out of line.[44] Nor are such tangled issues, endemic in late twentieth- and early twenty-first-century politics of knowledge, primarily about individual conformity or aberration. They invoke questions about the preservation of public trust and the creation of responsible epistemic citizenship and community: concerns notably absent from putatively universal, individualistic *a priori* theories of knowledge and action. They show how knowledge claims advanced and substantiated even by accredited, authoritative knowers are vulnerable to undermining by such extra-epistemological factors as social-political-ecological location; gender and race; ideology; and intransigent, willed ignorance, both individual and communal. They pose urgent questions about open and closed research practices and communities, where openness can influence the availability of funding and of other vitally necessary infrastructural support. They urge reexamining relations of trust among producers and

---

43. My thanks to Jamie Robertson for clarifying some of these issues.
44. Michel Foucault, *Madness and Civilization: A History of Insanity in the Age of Reason*, trans. Richard Howard (New York: Vintage, 1965), p. 58.

communicators of knowledge, reevaluating structures of integrity and trustworthiness within epistemic communities where "ideology" (commonly, but not always appropriately, construed as negative) and "fact" are often pitted against one another. Such struggles incorporate convictions about the "reduced" epistemic authority of putative knowers/teachers/writers who are Other than white middle-class, neither too young nor too old, heterosexual men. These self-sufficient "individuals" are tacitly assumed to be the rightful, paradigmatic occupants of such positions of pedagogical authority, whose modes of conduct exemplify pedagogical "excellence" at its best.

Acknowledging these difficulties opens the way to posing questions about who is appropriately positioned to do the reexamining, the reevaluating, given the uneven distribution of power and authority in epistemic/educational communities, where expertise is interwoven not just with financial concerns, but with such extra-epistemic factors as academic freedom in institutions of knowledge production and in the societies that house and support them. Thus, rather than standing as self-sufficient *individual* pedagogical exemplars, they need—with institutional-social support—to work toward renewed (collective) self-conceptions, supported by and supporting reconstructed science policies/knowledge policies and practices, capable of promoting democratic community and fostering epistemic justice. Democratically reconstructed policies and institutions could open deliberative spaces for engaging with ecological issues animated by the politics of knowledge and public trust, for exposing how an *instituted social imaginary* of self-sufficient individualism negatively infuses public perceptions and evaluations of scientific-epistemic practice in professional and more secular knowledge domains, an instituted imaginary that can as effectively endorse damaging practices as it can foster benign, ameliorative challenge and change. In these early days of sustained public debate, the situation is reminiscent of earlier feminist and antiracist consciousness-raising struggles, whose ongoing endeavors may produce resistance and despair, but whose multiple successes, however fragile, can generate a cautious optimism, tempered by a healthy skepticism.

This last thought returns to Medina's questions. The political-ethical failures of "society" Medina names herald the urgency of disrupting an apparently seamless epistemic imaginary: disruptions that can be animated by "individual" inspirations and advocacy, although these are rarely "individual" in the self-enclosed sense captured in the (mythical) figure of autonomous man. If it is plausible, as I propose it is, to allow that discovery requires a mind prepared, it is implausible to believe that such preparedness could be a purely individual effort. Feminist and other post-positivist epistemologists have exposed the limits of epistemic individualism. They have opened discursive spaces for learning to recognize and refuse to participate in social ignorance across a range of epistemically irresponsible practices that are reciprocally social-political and individual in their origins, enactments, and consequences. Perversely, then, they have encouraged a just measure of *strategic* skepticism, contrasted with the malign skepticism of the deniers. It is what all of us require in order to address and evaluate mass manipulation posing as "information."

In the current global climate crisis, as no reader of this chapter will need reminding, ecological responsibilities are paramount. They are epistemological, ethical, social-political, and inextricably intertwined with issues of social justice, across diverse populations and situations. They vary according to divisions of power and privilege as these are diversely inflected by gender, race, class, and geography, to name only some of the modalities. Epistemologically, they require knowing the specificities of situations and subjectivities in ways that defy generalizing from place to place; and they promote values that may also vary across situational differences. Fulfilling the epistemic responsibilities fundamental to engaging well with these issues calls for delicately crafted "situated knowledges" whose "reach" and/or pertinence from situation to situation can rarely be taken for granted.

Although questions about global warming, climate change, and ecological degradation are in the public eye, philosophical and more widespread analyses of their meanings and implications often represent them as matters of moral and/or political irresponsibility which call for individual self-restraint.[45] They are all of these and more. Yet, despite the urgency of the debates and the seemingly uncontroversial demonstrations of irrevocable damage ecological violations produce, many people in the educated, affluent West-North still cherish the comforting assumption that there is nothing to worry about; no need to promote way-of-life-changing measures. Such complacency betrays a conviction that "the world" has the resources to heal itself: it has done so in the past and, if "we" are patient and trusting, it will do so in the future. For ecological thinkers, this wait-and-see approach is naïvely dangerous. It relies on

---

45. See Stephen Gardiner, *A Perfect Moral Storm: The Ethical Tragedy of Climate Change* (Oxford: Oxford University Press, 2011).

selective assumptions about what "science tells us," neutrally, and with no genealogical inquiry into the (evolving) power-knowledge nexus that upholds the putative veracity of such voiceless, placeless, "expert" science. Although an overarching theoretical-scientific framework must urgently be developed, the issues need, equally urgently, to be addressed locally, situationally, taking the specificities and vulnerabilities of place, power structures, and populations into account. Hence, my aim is to reexamine climate change skepticism and doubt through a lens focused on entrenched, often contestable assumptions about how such issues can responsibly be *known*.

As readers of these words will be aware, harms consequent upon smoking, waste pollution, and global warming, encouraged and condoned by such doubts, are analyzed, emblematically, in Oreskes and Conway's *Merchants of Doubt*.[46] Less often acknowledged is the fact that these harms attest to deep-seated *ontological* assumptions about who we in the white affluent West are; about how we, singly and collectively, understand our place in the world. Their variable significance for feminist, class-specific, antiracist, and other vital areas of theory and practice remains under-examined in aggregated analyses, even though it is in such areas that some of the most urgent interventions are required. Yet when skeptics invoke a margin of scientific uncertainty to justify failures to regulate behavior, they often do so in conjunction with defending a conception of freedom—Liberty—assumed uncontestably to be theirs, individually and universally. Such assumptions invoke ethical-political questions about epistemic responsibility and entitlement. More fundamentally, they invoke *ontological* questions, not just about who "we" are, but about how we can live responsibly together in the human, natural, and social-political world, and respectfully of that world.

My thinking here is continuous with the manifesto with which I conclude the introduction to *Ecological Thinking*:

> [T]hinking ecologically carries with it a large measure of responsibility—to know somehow more *carefully* than single surface readings can allow. It might seem difficult to imagine how it could translate into wider issues of citizenship and

---

46. Naomi Oreskes and Eric M. Conway, *Merchants of Doubt: How a Handful of Scientists Obscured the Truth on Issues from Tobacco Smoke to Global Warming* (New York: Bloomsbury Press, 2010).

politics, but the answer, at once simple and profound, is that ecological thinking is about imagining, crafting, articulating, endeavoring to enact principles of ideal cohabitation.[47]

That observation brings social epistemology and ethical theory together in relation to feminist and antiracist theory, and to ecological, ethical, and political issues: they pick up the threads at places where these lines of inquiry intersect. This conceptual frame extends into matters of vulnerability, incredulity, ignorance, and trust: modalities also unevenly distributed across populations of internal diversity and diverse relations to the wider world. Hence, social-justice issues cannot be investigated in one-size-fits-all analyses: they must strike a balance across particularity, multiplicity, and commonality. Engaging with such considerations, I suggest, leads into questions about education, human rights and the environment, the politics of care in techno-science, and the public cultivation of ignorance and doubt in climate change skepticism that plays into human and ecological vulnerability while fostering an atmosphere of complacency, incredulity, doubt, and distrust.

Most urgently, as I have noted, these ideas pose fundamental questions about who we think we are, we who have made the world for which we are now accountable and who need urgently to develop ways of restoring and preserving that world for others who come after us: other generations, cultures, and races who must live with the effects of our practices. As I note in the introduction, I want to speak and hear the question "Who do we think we are?" provocatively, normatively: not as a rhetorical question, but as a political-ontological challenge such as might be posed in situations of surprised dismay at presumptuous behaviors, intrusive or offensive assumptions. It is in thinking of this way of posing the question that I name social justice as an overarching issue, as ethical-epistemological-political, and as articulated from a position shaped by feminist, anti-racist, post-colonial, and other Others' engagement in recognizing that knowing responsibly is *sine qua non* for responsible social action. It is vital to engaging with people, places, things, and situations "on their own terms," where "responsible" includes recognition and responsiveness, as far as this is possible. The addressive formulation of

---

47. Lorraine Code, *Ecological Thinking: The Politics of Epistemic Location* (New York: Oxford University Press, 2006), p. 24.

the question emphasizes its necessarily interactive character: it challenges tacitly or overtly individualistic ontological presuppositions.

Contentiously, in a 2013 post to the *Feminist Philosophers* list, Sally Haslanger maintains: "philosophers don't think about social structures very clearly. In the normative domain, they tend to be more focused on the individual—what makes an individual's actions right or wrong. Or they think about the state—when is a state legitimate and what is the state obligated to do for its citizens."[48] This observation, I suggest, pertains principally within the confines of analytic orthodoxy where "individuals" indeed are starkly "individuated" and often, in consequence, supremely self-interested. One of my goals, by contrast, is to generate a subversive rethinking of (often tacit) Western ontological convictions about individualism and universal human sameness, which—among other contestable practices—underpin "our" presumptive entitlement to consume and pollute, as we will. This goal is meant to prompt us, as educators and activists, to engage in projects designed to generate a shift in "the ontology of the self"; hence in the policies and practices governing quotidian and public-political social structures in the white affluent Western world. One large piece of it would be to instill a measure of caring that may seem remote from *bona fide* questions about knowing, or polluting, but is germane to answering them well. This claim is, perhaps, one of the hardest to defend, and the most vital to enact.

A central contrast is with practices of super-imposing categories and explanations that subsume differences under pre-formed categories, with inadequate attention to whether an appropriate "fit" can be achieved: a perhaps inadvertent practice, endemic in the background ontological assumptions that inform the epistemologies and ethical theories and practices of the Anglo-American mainstream. Hence, as I note in the Introduction, in George Yancy's volume *The Center Must Not Hold*, a pivotal thought, variously articulated, is this issue of who "we" think we are in taking for granted our entitlement to live as we do.[49] It is implicit in Susan Babbitt's analysis of the ineluctable "whiteness" of the philosophy taught in most English-speaking universities, with its multiple

---

48. *Feminist philosophers*: "What is natural and what is social?" Sally Haslanger: Posted: 22 Feb 2013 10:41 AM PST.
49. George Yancy, ed., *The Center Must Not Hold: White Women Philosophers on the Whiteness of Philosophy* (Lanham, MD: Lexington, 2010).

exclusionary effects; in Alexis Shotwell's exploration of the effects of "whiteness as method" in sustaining diverse oppressions; and in Shannon Sullivan's thoughts about how philosophy's secularity works to discourage people of color from participating in the discipline.

Recalling the persuasive conceptual framing from Robert Figueroa and Gordon Waitt, also cited in the introduction, I am affirming that openness to the potential epistemic legitimacy of "affect, materiality, creativity, and multiplicity" allows for capacious ways of recognizing and acknowledging subjectivities that differ—even radically—from "one's own," where "[n]ormative elements of moral terrains include embodied narratives about what behaviours are permissible, who belongs where, how we perceive the moral status of other bodies (human and non-human), and the ways in which we establish moral, social, and political identities in embodied relations to space and place."[50] Their reference is to indigenous Australians' urging would-be climbers to respect the sacredness of Uluru in ways that require them not just to *know* differently, but to *be* different from how they have assumed they can be in regarding the natural world as up for grabs—or for climbing!—at anyone's whim. *Knowing differently* requires learning to know Uluru as a sacred site that is not theirs to appropriate, where climbing counts as appropriation in ways that might not pertain elsewhere. It calls for a reconceived ontology and an ethics of place. It requires a thoughtful move away from convictions of autonomous self-certainty and entitlement, working toward achieving respect for an otherness that is not theirs to obliterate or to own, whose standing is not theirs to define or to dispute. The young man who insists "even though they say you are not supposed to climb it. I climbed it anyway. But, that's just the way I am. I'm not going to not climb it" has much to think about: he needs to learn the limits of self-certainty.

How these examples bear on larger ecological aims of developing principles of ideal cohabitation may be puzzling. But eschewing the radical individualism of Western philosophy and social theory and endeavoring to think communally, socially, co-operatively and across human and situational differences is germane to understanding ecological harms in their specificity and generality. Their impacts, effects, and potential remedies cannot, again, responsibly be analyzed or enacted in one-size-

---

50. Robert M. Figueroa and Gordon Waitt, "Climb: Restorative Justice, Environmental Heritage, and the Moral Terrains of Uluru-Kata National Park," *Environmental Philosophy* 7, no. 2 (Fall 2010), pp. 135–63.

fits-all approaches. Contesting the tenacity of the center simultaneously contests the suitability, applicability, and pertinence of situation-insensitive diagnoses, policies, and remedial measures.

As I have been urging, engaging knowledgeably with these matters requires rethinking, reenacting *who we are*, in ways sufficiently sensitive and powerful to dislodge such sedimented convictions. This is the hardest, most urgent demand: it is easier, more imaginable, to think and participate on the surface, so to speak, in revisionary ways of doing, thinking, knowing. But to practice a philosophy and develop a pedagogical practice which requires—*must* require—such rethinking and reenacting is ontologically-epistemologically radical, upheld as those assumptions are by the instituted social-political-epistemic imaginary in which we inhabitants of the affluent West live and think have our being, however obliquely or contrarily. Taking these matters seriously calls for crafting and living critically renewed conceptions of human subjectivity and agency at a social, collective level. These requirements are analogous to late twentieth-century consciousness-raising practices in calling for critical genealogical analyses of who we are and how we are accustomed to live, conducted in ways sufficiently discerning to unsettle many of the expectations that inform our everyday thoughts and actions. It is hard work to engage with this requirement; it is easier to devise revisionary ways of doing, thinking, and knowing. But to practice a philosophy that requires unsettling and reenacting basic assumptions about *who we are* is ontologically and epistemologically radical, upheld as these are by an instituted social-political-epistemic imaginary (following Cornelius Castoriadis) in which as inhabitants of the affluent West we live and think, however obliquely or contrarily.[51] It is in this respect that we, as educators, carry a special responsibility to know ecological issues carefully so that such knowledge can inform our pedagogical practices—a responsibility that manifests, also, in generating an intelligent scepticism in our students and colleagues, about the taken-for-granted rightness of who "we" are and how "we" live.

Oreskes and Conway's *Merchants of Doubt* speaks to readers across a spectrum from an educated, concerned public, to environmental and other scholars and to epistemologists. Significant is its focus on questions

---

51. See for example Cornelius Castoriadis, "Radical Imagination and the Social Instituting Imaginary," in Gillian Robinson and John Rundell, eds. *Rethinking Imagination: Culture and Creativity* (London: Routledge, 1994), pp. 136–54.

of knowledge and ignorance, where class and gender figure centrally. While race-related questions are underaddressed, the conceptual apparatus, I believe, readily expands to accommodate them. Strategies for keeping public ignorance alive and active are meticulously analyzed in the text. It meshes with work on *agnotology* in its diagnostic analyses of areas of ignorance that call out for investigation analogous to the analyses epistemologists devote to discerning what makes knowledge possible.[52] Diversely, within such areas of investigation as the effects of tobacco smoke, the environmental and human health impacts of DDT, and the implications of widespread failures to control greenhouse gas emissions, these studies examine the power of uncertainty to persuade people that the current state of knowledge is insufficient to justify modifying everyday practices.

Kristin Shrader-Frechette's investigations of human and environmental damage contribute immeasurably to counteracting such ignorance, especially in the delicate balance she achieves between objective scientific evidence and specifically detailed environmental harms with their ethical-political and indeed ontological implications for particular examples of human suffering.[53] Her thinking also facilitates extending this project into current engagements with the politics of care in techno-science. As I have noted, social epistemology's conceptual shift away from practices of uttering knowledge claims as though into a void and toward the language of speakers and hearers, where credibility and trust become central is germane to the pedagogical significance of these observations. Shrader-Frechette exposes the importance of individual and social-communal uptake for testimony's being heard, understood, contested, and/or acted upon, in communities where power-privilege differences enhance or inhibit its receiving a fair, just hearing. Testimony is crucial to ecological thinking with its reliance on experts and putative experts, given that nonexperts often have little choice but to

---

52. The reference is to Robert N. Proctor and Londa Schiebinger, eds., *Agnotology: The Making and Unmaking of Ignorance* (Stanford: Stanford University Press, 2008).

53. Kristin Shrader-Frechette, *Taking Action, Saving Lives: Our Duties to Protect Environmental and Public Health* (New York: Oxford University Press, 2007); and Kristin Shrader-Frechette, *Environmental Justice: Creating Equality, Reclaiming Democracy* (New York: Oxford University Press, 2002). See Iris Marion Young, *Inclusion and Democracy* (Oxford: Oxford University Press, 2000), p. 39; and "Activist Challenges to Deliberative Democracy," *Political Theory* 29, no. 5 (Oct. 2001), pp. 670–90. Shrader-Frechette notes Young's influence on the ethical-political framing of her research.

trust "expert" testimony. Yet the politics of testimony are notoriously fraught, drawing ecological-environmental issues into a frame where questions about *particularity* also occupy a central place: such questions as "How can we know well enough to respond well to the effects of the current crisis for these—specific—issues?" They are challenging because standard theories of empirical knowledge with a focus on "knowledge in general" have little patience with particulars, specificities. The risk of descending into "particularism" and being dismissed accordingly haunts such projects. Social epistemology, with its emphasis on testimony, makes space for questions of this sort, showing that well-narrated analyses can illuminate complex social-political questions. It is innovative in exposing the uneven effects of "manufactured uncertainty" for diverse populations, especially those not represented in the figure of "autonomous man"; for reclaiming advocacy as a responsible epistemic practice; for its attention to ignorance and skepticism; and for its pertinence to educational theory and practice.

Pedagogically, engaging with particularity has immense value because careful teachers and mentors can work with it to generate a just measure of strategic skepticism that students—and all of us—require in this age of mass manipulation by "information." Observing the difficulties, for the "haves" of imagining how it is *not* to have affirms the urgency, for educational practice, of educating our imaginations and the imaginations of our students and colleagues to stretch beyond the familiar, to acknowledge and respect strangeness, and to recognize the caution required in treading respectfully on unimaginable territory, in assuming to know too much from too narrowly imagining. Matters of social justice cannot be addressed in one-size-fits-all analyses; hence my emphasis on the ecological implications of lived particularity and difference.[54]

---

54. I borrow some of these reflections from my "Particularity, Epistemic Responsibility, and the Ecological Imaginary," *Philosophy of Education Archive* (2010), pp. 23–34.

Chapter 2

Doubt and Denial

Epistemic Responsibility Meets Climate Change Skepticism

When Virginia Woolf, in *A Room of One's Own*, deplores her exclusion from the carefully manicured grounds and hallowed halls of an Oxbridge college and its library, she has no illusion about the reasons why. She is a woman. "Only the Fellows and Scholars are allowed here [on the turf]: the gravel is the place for me," she writes. Small wonder that she comments of her fictional, authorial self: "'I' is only a convenient term for somebody who has no real being,"[1] for its pertinence extends beyond the putatively anonymous authorship of this text. The exclusionary issue is Woolf's sexed/sexuate being. No doubt had she been otherwise "other"—not upper middle class, not white—she would have been doubly, triply excluded. But being a woman was enough. No such multiple identities were at issue, for in the actions and policies that kept her out, sexual difference overrides and sustains other differences even in their diverse modalities, affirming the outrageousness, the unthinkable invasion of a woman in the halls of academe. Equally compelling is the idea that the "gravel" is the place for her, which is literally and metaphorically far-reaching in its implications. This reference to place, to a lowly and uneasy place at that, presages the views germane to the issues I discuss here, in this chapter, where place, situations, spaces are

---

1. Virginia Woolf, *A Room of One's Own* (London: Granada, 1977 [orig. 1929]), pp. 6, 8.

ecologically, epistemologically, and ontologically significant in ways their merely background taken-for-grantedness effectively obscures.

The minimal significance mainstream Anglo-American theories of knowledge and ethics have accorded to place, in which they are supported by and support a picture of knowers and/or "moral agents" as mere place-holders in/on unspecified and putatively irrelevant places, upholds long-standing assumptions about the interchangeability of knowers and "the known" in "properly objective," affect-free acts of knowing, doing, and being. The very idea that knowing could be constitutive of place and place of knowing in ongoing reciprocal practices appears, in the postpositivist legacy, to threaten epistemic anarchy or to initiate a descent into the chaos of relativistic unknowing. Let us think, then, about the gravel: the gravel as "the place for me." In the scenario Woolf constructs, the gravel would not be the place for "him" or for his *sembables*, whose rightful place need not be specified because it is everywhere and anywhere: that the turf with its well-tended comfort is his, is presupposed in institutional and social structures so firmly in place that they need neither be noticed nor spoken. In such unmarked places, hegemonic knowing and doing occur: they work to determine the structures and demographic hierarchies enacted on/with those relegated to "the gravel." (It does need to be added, though, that not every man in Woolf's England would be justified in claiming a place "for him" on the higher ground. Class-specific, race-specific, age-specific matters do not figure in this brief allusion.)

As a way of understanding "how sexual differences can aid our responsibility for nurturing the sustainable ecologies of our local and global communities, environments, and interactions . . . [and their place in thinking about] the impact and expressions of social justice and citizenship,"[2] I am proposing certain links between Luce Irigaray's observation that "[t]o construct only in order to construct nevertheless does not suffice for dwelling," and my thematic claim that "ecological thinking is about imagining, crafting, articulating, endeavoring to enact principles of ideal cohabitation."[3] In tracing some lines of thought that bear, if obliquely, on the resources ecological thinking offers for enacting a

---

2. Quotation from the description of the panel "Sexuate Sustainable Practices and Ecologies" at the conference "Sexuate Subjects: Politics, Poetics, Ethics" (University College, London, December 2010), where I presented an earlier version of this chapter.

3. Luce Irigaray, *The Way of Love* (New York: Continuum Press, 2004), p. 144; Lorraine Code, *Ecological Thinking: The Politics of Epistemic Location* (New York: Oxford University Press, 2006), p. 24.

viable conception of *sexuate* subjectivity, I am indebted to Alison Stone's reading of Irigaray who, she says, uses "sexuate" in three principal senses: (1) Of law and rights; (2) Of a culture insofar as it recognizes sexual difference; and (3) Of our nature *qua* sexed.[4] I draw, if sometimes tacitly, on the second and third senses, attempting to make good the promise implicit in the above quotation from *Ecological Thinking* by considering some implications for sexuate subjects of thinking ecologically, and for ecological thinking of addressing the sexuate being of ecological subjects. These ideas bear on a larger unease with climate change skepticism and the mockery it performs, in the name of abstract instrumental reason, on such nuanced idea(l)s as "dwelling" and "ideal cohabitation."

A powerful force of resistance to acknowledging the ecological implications of climate change is a stubborn commitment, in late twentieth- and early twenty-first-century Western/Northern capitalist societies and in mainstream Anglo-American philosophy, to a sovereign individualism: to a commitment manifested by knowers and doers who, it seems, have no sex, no place, and no "personal" allegiances in any ontological, self-constituting, or self-sustaining way. The "god trick" Donna Haraway deplores continues to exert a strong, if elusive, appeal.[5] As I note in *Ecological Thinking*, ethical self-mastery, political mastery over unruly and aberrant Others, and epistemic mastery over the "external world" persist as the goals of the Enlightenment legacy, in philosophy and in the social-political imaginaries of (mostly white) capitalist societies. These discourses shape and are shaped by an individualist reductivism for which epistemic and moral agents are undifferentiated subjects, isolated units on an indifferent landscape, to which their relation is one of disengaged indifference. They enlist ready-made, easily applied categories to control and contain the personal, social, and physical-natural world within a neatly manageable array of "kinds," obliterating diversity in projects of assembling the confusion of those worlds into maximally homogeneous units. Such taxonomies sustain the *instituted* social imaginary[6] for which instrumental rationality is the epitome of human reason, while

---

4. Alison Stone, *Luce Irigaray and the Philosophy of Sexual Difference* (New York: Cambridge University Press, 2006), p. 16n.

5. See Donna J. Haraway, "Situated Knowledges," in her *Simians, Cyborgs, and Women: The Reinvention of Nature* (New York: Routledge, 1991), pp. 183–202, at p. 189.

6. I use the term "imaginary" in Cornelius Castoriadis's sense. See his "Radical Imagination and the Social Instituting Imaginary," in Robinson and Rundell, eds., *Rethinking Imagination: Culture and Creativity* (London: Routledge, 1994), pp. 136–54.

the *instituting* ecological imaginary I have endeavored to configure and promote seeks to disrupt these assumptions, all the way down. Ecological subjectivity, *sexuate* subjectivity animates these contestations, struggling to dislodge the intransigent, overweening power of mastery and domination. Here, then, a focus on sexuate subjectivity works to uncover, and to demonstrate, how sexual difference has played into—has, in effect, been in many respects constitutive of—these oppressive and damaging structures; how bringing it into central focus could perform a version of consciousness raising at a deeply theoretical level, where the effects of sexual differences might be exposed and reevaluated on a slate wiped relatively clean of the residual baggage they carry and the detritus they leave in their wake. The hope—perhaps the impossible dream—is that ecological thinking could reconfigure subjectivity, sociality, knowledge, and moral-political thinking by enlisting the potential of sexuate epistemic and ethicopolitical practices to produce *habitats* where people can live well together, and respectfully with and within the physical-natural world.

These grand gestures of contestation cannot be undertaken only as a large, overarching project, even though they have large and grand goals. They need equally to be addressed and analyzed locally, not just in being located in specific places, although this matters too, but also in their diverse particularities, in which even when the principal focus is on sexual difference and sexuate subjectivities, these will be enacted and weighed differently in ecologically specific locations where the detail of situation and place will be as significant as will the connections and differences between them. Hence, this line of thought involves a cautious, even an ambivalent turn toward particularity/particularities: knowing, acting, thinking ecologically involves intelligently mapping, and caring about, particularities while guarding against initiating a descent into pure particularism.[7] It thus requires a delicate balancing: thinking by analogy from place to place, maintaining an openness to learning from disanalogy when analogies fail to go through or prove unstable.

Climate change skepticism and its advocates stand in a coldly antithetical relationship to ecological thinking with its implicit commitment to sustainable practices and equally to the assumptions about sexuate subjectivities and the politics and ethics of knowledge that (sometimes

---

7. See in this regard Lorraine Code "Particularity, Epistemic Responsibility, and the Ecological Imaginary," *Philosophy of Education Archive* (2010), p. 23–34.

tacitly) inform it. Climate change skepticism, with the *agnotology* it generates, trades on a certain cynicism about knowledge, about human (anonymous but instrumental) subjectivity, about the earth/the world (in Hannah Arendt's sense) and about people's lives. It is animated by and animates the starkly atomistic, resolutely neutral conception of subjectivity, in practices and rhetorics that oppose sustainable ecological regulatory measures from a (misbegotten) conviction that they—almost by definition—tamper with, and indeed jeopardize, said freedom.

A second, interconnected focus of discussion in this chapter is Naomi Oreskes's and Eric M. Conway's 2010 book *Merchants* of *Doubt*, which I have begun to read in chapter one for its astute analyses of certain major obstacles in the way to fulfilling "our responsibilities for nurturing the sustainable ecologies of our local and global communities, environments, and interactions," thereby endeavoring to foster social justice. I take the Oreskes and Conway book as my second point of entry into these thoughts because of how clearly it depicts the quintessentially neutral stance adopted by "science," at least in the public eye: science intentionally homogenized and reified in—dispassionately—sustaining systematic cultivations of doubt and indeed of ignorance, in a dominant, still predominately white Western society and population; a land of politically-invested expansions of the margins of uncertainty in some academic and much public discussion of the implications of climate change. Exemplary are the book's analyses of the ethical-political implications of manufactured doubting: albeit sometimes tacitly, it proposes ways of thinking well, ecologically, in attempting to counter such skepticism.

Three themes are central to the next sections of this chapter, whose purpose is to evaluate certain insights that emerge from exploring their interconnections. First, I read a face-off between ecological thinkers-activists and climatologists on one side, and climate change skeptics/deniers on the other, by way of showing how questions pivotal to the epistemology and politics of testimony bear, directly and indirectly, on debates surrounding climate crisis, gender justice, and the politics and ethics of development projects. In so doing I refer to "ecological thinkers," intending thereby to name theorists and activists who work toward understanding and generating a widespread activism, committed to countering or mitigating climate change from ecologically-informed positions. Here, thinking ecologically encompasses wide-ranging, engaged, interpretive investigations of the implications and effects of historical-political-geographical-intellectual "situatedness" for citizenship and politics, broadly conceived. It initiates and seeks to nourish a widespread activism manifested, for example, in town hall meetings, pedagogical settings, newspapers, television, and publicly available articles and essays: as a social movement, it is deeply reliant on responsibly achieved and promulgated knowledge. It engages critically and constructively with the "manufactured uncertainty" that works to limit such investigations. The issue, as I note above, is about critically imagining, crafting, and endeavoring to enact principles of ecologically sound human cohabitation with one another, and in and with the wider world, both animate and inanimate. This point of entry derives from a stark, and troubling, contrast between climate change skeptics and ecological activists, yet it does not assume that such a contrast alone is definitive of the issues under discussion. Still, framing the questions in this way facilitates a plausible direction for approaching the specificities and generalities of contentious, multiply tangled, diversely situated and populated questions pertinent to ontological-epistemological questions about "our" epistemic and ethical relations to and practices within the world.

Secondly, in a stubborn stand-off across widely differing positions in respect to issues of this kind, intelligent debate risks descending into a contest—to fighting science with science (thereby tacitly acknowledging that "science" in its substance is a human construct—neither a divine revelation nor a univocal reading of the world). For non-scientists, it risks descending into a somewhat different contest between/among conflicting expert testimonies singly and collectively articulated, in which decisions about where, reasonably, to place belief and trust are ethically-politically and epistemologically urgent, yet also fraught. In consequence, reliance on testimony generates fundamental questions about human subjectivity, agency, and freedom that have long been relegated to the non-rational margins of *bona fide* theories of knowledge. Climate change-skeptical positions tend, if often tacitly, to presuppose an abstract, impersonal conception of autonomous gender/race/class-neutral subjectivity which, despite the neutrality of its generic presentation is inevitably gendered, raced, and otherwise socially-politically-economically inflected, and situated in philosophically-epistemologically significant ways. The implications of such situatedness require closer scrutiny, frequently case-by-case.

Third, to illustrate how these lines of thought might come together in taking *gendered* subjectivity into account, I revisit Rachel Carson's ongoing *ad feminam* treatment by the science skeptics of her time, and still—albeit differently—in ours. Her 'case,' and the contest of testimonies it continues—if indirectly—to generate, exposes the sexed-gendered specificities concealed, in her time and in ours, beneath the façade of the presumptively generic epistemic and moral subject.

"Manufactured Uncertainty"—Merchants of Doubt

Climate change skeptics stand in a starkly oppositional relationship to ecological thinking,[8] to larger commitments to fostering and maintaining ecological sustainability, and to presuppositions about subjectivity and the ethics and politics of knowledge that inform and shape these modes of thought, together with the actions they inform. As I note earlier in this chapter, I contend that such skepticism trades upon a certain cynicism about knowledge, the earth/the world, and people's lives, in the name of a profit-driven yet also curiously conceived *freedom*-promoting and -preserving instrumentalism. "Cynicism" as I invoke it here captures a generalized, if often tacit, disdain: a distrust of settled and/or of innovative ethical and social values in mass society, especially where high expectations of the occupants of positions of authority, and of the integrity of public institutions, tend to go unfulfilled. It manifests in a generalized disillusionment and distrust of putatively ameliorative social-political initiatives and organizations: in a sneering fault-finding.

Analogous to such thinking about skepticism are Inmaculada de Melo-Martín's and Kristen Intemann's 2018 views about "normatively inappropriate dissent" (NID), articulated in *The Fight Against Doubt*.[9] In this rich text, a point of emphasis significant to my analyses here is the attention the authors pay to a cluster of contrasts between matters of *epistemic* trust and normatively inappropriate dissent with respect to scientific knowledge claims/contentions. Melo-Martín and Intemann are reluctant to articulate precise characteristics that define what makes dissent normatively inappropriate. Their focus is instead on the effects of NID, on its deleterious influence on scientific progress. Also pertinent is a putative contrast they draw between scientific knowledge claims and *facts*. This subtle and elusive distinction is pivotal to deliberations

---

8. See Code, *Ecological Thinking*.
9. Inmaculada de Melo-Martín and Kristen Intemann, *The Fight against Doubt: How to Bridge the Gap between Scientists and the Public* (Oxford: Oxford University Press, 2018).

about how, where, and by what criteria it is plausible to invoke appeals to trust in reliance on—in appeals to—verbal (= oral, visual, or written) communications of both scientific and "secular" knowledge matters.

The form of skepticism with which I take issue is a manifestation of normatively inappropriate dissent in the sense that those who espouse it fail to disentangle factual and ideological aspects from one another. This form of skepticism is commonly animated by and itself animates the starkly atomistic, self-contained, and affectively detached conception of human subjectivity I find problematic. It is a conception of subjectivity that involves practices and rhetorics that, in their very substance, oppose ecologically informed social-political regulatory measures enacted to evaluate and to limit climate change, usually from a conviction that these are measures designed to tamper with, indeed, to jeopardize individual human freedom. Such freedom is commonly cast as an incontestable human entitlement and goal. Questions about *who* the "beneficiaries" of such refusals are, and why they have come to be the chosen ones, rarely enter these debates.

This chapter's second point of departure is a single text: Naomi Oreskes and Erik M. Conway's 2010 *Merchants of Doubt*, to which I refer in the introduction and again in chapter 1. The book is pivotal to these discussions for its engaged, catalytic analyses of scientific-epistemological questions in ways that speak to specialists and to an informed or would-be informed "general public" alike. It is likewise pertinent for its "situated" analyses (again following Haraway) within extensive yet subtle critiques of science-oriented epistemological practices, current in present-day capitalist free market societies as these involve ordinary folk, highly qualified scientists, and an entire range of populations and practices, situated among and between them. Focusing on Oreskes and Conway's book is particularly appropriate as a point of entry into this inquiry in consequence of its path-breaking approach to matters at issue through the history of Western-Northern science, ethics, epistemology, and politics, all the while foregrounding, and condemning the implications of present-day American exceptionalism for engaging just such matters. For twenty-first-century Anglo-American epistemology and politics of knowledge, the book was—and is—startling in its exposure of a curiously delusionary insistence by deniers—doubters—among US scientists, and then-politically invested others, on assigning equal weight to "both sides" of climate change debates before approving regulatory measures. Oreskes and Conway contend—plausibly—that the "sides" do not have equal claim to credibility, epistemic responsibility (a dominant issue here), or scientific-intellectual integrity. Hence the urgency of the need, at once, to understand and address the science and to consider the weight of ideological opposition masked as a commitment to fairness and equality—dressed up in a uniquely American fear of governmental "interference" with a quintessential US value: Liberty.

Sparked by Frederick Seitz's contention that focusing on global warming was simply a scare tactic, a consequence of pandering to irrational fears of environmental calamity by scientists seeking fame and fortune, the authors document diverse effects of a quintessentially *neutral* cognitive stance that trades on a conception of "science," homogenized and reified. That position dispassionately sustains a level of doubt, of uncertainty, and indeed of ignorance in a dominant, predominately white affluent Western society and population. For this populace, the claim "science has shown" carries a cachet enhanced by such presumed neutrality, which suggests that the facts alone are so compelling that reason cannot withhold assent. Yet the veil of neutrality behind which this impersonal claim is commonly articulated screens out any thoughts that specifically situated, and not disinterested, human subjects have produced the "science" invoked, or that the effects of what "science has shown" will be enacted in specifically situated, thence not disinterested or homogeneously participating, lives. Such inquiry works from a taken-for-granted presumption of human sameness in—or in respect to—matters epistemic, ethical, ecological. No articulated conception of epistemic or moral-political subjectivity informs the inquiry; a presupposition prevails that the knowledge discovered or found will fall on neutral ears and uptake will be uniform and equivalently neutral. The epistemic practices the authors depict are coldly inhospitable to examining how subjectivities and situations are diversely constituted and enacted in this bleak picture of a social-ecological order where it is hard to see how anyone could be "at home."

That many climate change deniers are motivated by financial gain will come as no surprise. More striking, philosophically and practically, are their fears about the *loss of freedom*, generically conceived, that ecologically informed regulatory measures will presumably entail. According to Oreskes and Conway, such fears tend even to override financial concerns in significance, in the twenty-first-century USA and (with local variations) throughout the affluent populations of the white Western/Northern world. An ongoing concern (albeit, again, with regional and demographic variations) is that if threats to "the environment" require regulatory measures, then they will interfere intolerably with human freedom—with Liberty—hence cautionary interventions should not be contemplated. The conception of freedom operative here is at the core of the autonomous self-contained and self-sustaining individualism that infuses the liberal democracies of the white capitalist Western-Northern world, whose masculinist-patriarchal enactments and effects—its demeanour of neutrality notwithstanding—have long been a focus of feminist critique. In a free society, the argument goes, "we" will—*ex hypothesi*—refuse to be constrained in our freedom to smoke, to pollute, to use pesticides when and as we will. This refusal attests to a governing conception of subjectivity according to which, generically and neutrally, there can be no contestation of the freedom at issue, no question about its universal attainability, its everyday observance, and its viability as a sacrosanct human value. There is no interest in who these perfectly free subjects are or in what ways sexed, gendered, racial, class-and-age-specific or any other differences from a white sex-gender neutral norm could have played a constitutive part in shaping the communities, environments, and interactions where such freedom is defended. The apparent explanatory power of an instituted social imaginary can be such that certain local differences, departures from an established norm, are rendered inconsequential, even invisible.

Operative in these considerations is a conception of freedom that is sharply individualistic—the freedom of the "buffered self"[10] of masculinist modernity—born of the imagined self-sufficiency integral to a frontier mentality. It is nourished not by an imaginary of mutuality and responsible dwelling, but of individual entitlement, acquisition, and instant gratification. Yet such freedom is unsustainable in itself, especially in looming crises of scarcity, and in economically-ecologically less affluently placed populations than its articulators presuppose. It is a freedom without the resources to sustain even the subject who adamantly affirms it as his.

According to Oreskes and Conway, a conviction has prevailed in post-1970s America that there can be "no freedom without capitalism and no capitalism without freedom."[11] When it gradually became apparent that industrial emissions were damaging human and ecosystem health,

---

10. Genevieve Lloyd, *Providence Lost* (Cambridge, MA: Harvard University Press, 2008), p. 322.

11. Oreskes and Conway, *Merchants of Doubt*, pp. 64–65.

and that "regulating" such emissions—which, allegedly, "flew in the face of the capitalist ideal"[12]—could be the only responsible response, the struggle turned into a battle of science against science. (It is worth noting that this idea of competing sciences does not square with the image of science as a neutral exercise of rational inquiry.) The solution among nay-sayers, as I will describe momentarily, was to denigrate this idealized view of science. Although this thought might seem to be consistent with my view, the similarity is superficial. Rather than engaging in specifically situated practices of investigation, the nay-sayers have opted for stubborn, global, ideologically motivated denial. Thus, against the (US) Environmental Protection Agency (EPA)'s efforts to expose harms produced by tobacco smoke, skeptics charged that such proposals merely allowed "bad science to become a poor excuse for enacting new laws and jeopardizing individual liberties; . . . It wasn't just money at stake, it was individual *liberty* . . . Today, smoking, tomorrow . . . who knew? By protecting smoking, we protected freedom."[13] In controversies over multiple forms of pollution the argument was advanced that if science was working against "the blessings of liberty . . . they would fight it as they would fight any enemy. For indeed . . . science *was* showing that certain liberties are not sustainable—like the liberty to pollute."[14] It is in this climate of fierce resistance to knowing/acknowledging immanent threats of climate disaster and its unpredictable consequences that the politics of testimony I discuss are being enacted, in what amounts to a struggle for ascendancy between an ecological imaginary and what Val Plumwood aptly characterizes as "an epistemology of mastery."[15] She is not alone in so doing.

As I have suggested, in the affluent twenty-first-century white Western-Northern world, a stubborn, systemic prejudice against disrupting the complacency of the status quo is embedded in widespread charges that climate change scientists are promulgating irrational fears. Such skepticism is powerful in its influences, especially when it is perpetuated by and insulated against condemnation by nay-sayers for whom there are

---

12. Oreskes and Conway, *Merchants of Doubt*, p. 65.
13. Oreskes and Conway, *Merchants of Doubt*, p. 145.
14. Oreskes and Conway, *Merchants of Doubt*, p. 239.
15. Val Plumwood, *Environmental Culture: The Ecological Crisis of Reason* (London: Routledge, 2001).

no immediate rewards for critically reexamining or contesting entrenched convictions that science-as-usual has all the right answers, or for critically re-evaluating everyday, ecologically irresponsible practices. For members of a comfortable (if again usually white) social-economic elite, acknowledging and acting to minimize the injustices climate change skepticism condones would entail significant personal "losses," not just of physical comfort, but of the myriad privileges and self-certainties that structure such peoples' entire ways of being—and rightly so, in the going view. They need take no measures to reduce consumption, to tread gently (whether literally or metaphorically), or to consume "more sustainably," more carefully: the world has always "healed" itself, and is bound to continue doing so. So goes the rhetoric.

Why would people whose lives are constructed around the illusions and comforts such unsustainable practices uphold be prepared to relinquish these privileges and ways of being that have long been theirs and to which they assume uncontested entitlement? This is the pressing question: it is at once epistemological, ethical, ontological, and political. Again, charges to the effect that climate change scientists are promoting "irrational fears" highlight the intensity of this oppositional struggle between those who think and act within a hegemonic imaginary of mastery, plenty, and entitlement and those who are committed to interrogating and destabilizing it. It is true that these latter commitments cannot be wholly effectual if/when they persist as "individual" efforts or projects; but again, neither can epistemic individualism itself continue to be accorded overarching value. Where the situation can be portrayed as being about fighting science with science, and where "good science" and "bad science" tend to be categorized less according to the quality of the knowledge—and the thinking—they produce; more according to the social-political interests they serve, questions about *responsible* epistemic practice seem not to figure. Despite their being germane to the matters at issue, they have found little traction in public debate.

Consider this now-classic example: into a 1990s debate about risks posed by Environmental Tobacco Smoke (ETS) in North America, a publication *Bad Science: A Resource Book* was inserted.[16] (Oreskes and Conway

---

16. Oreskes and Conway observe: "The phrases 'excessive regulation,' 'over-regulation,' and 'unnecessary regulation' were liberally sprinkled through the book. Many of the quotable quotes came from the Competitive Enterprise Institute (CEI), a think tank promoting 'free enterprise and limited government' and dedicated to the conviction

call it "a how-to handbook for fact fighters."[17]) Its avowed intention was to challenge the authority and integrity of the US Environmental Protection Agency's (the EPA's) research into the effects of second-hand tobacco smoke. In consequence, the contest is presented to the public as though it were a straightforward debate between adherents to two equivalent, commensurable lines of inquiry and sets of beliefs; yet where an awkward background assumption prevails that "science," cursorily aggregated, is nonetheless a source of certainty, and a "clean," neutral source at that. Such presuppositions notwithstanding, in the eyes of the naysayers, each "side" is driven by a specific, if tacit, agenda which many members even of a well-educated lay public are not qualified to judge; nor, were they/we to enter the debates, could we/they readily claim a hearing. Resistance to the idea of ideologically motivated science—science with sides—continues to bear traces of a residual positivism-derived refusal to allow that there could be more than one *way of knowing* specific subject matters, events, objects, theories. Descriptively and normatively, an epistemological assumption has (often tacitly) prevailed that knowledge worthy of the name must be univocal and definitive. This philosophical assumption is preserved in the impersonal pronouncement "Science has proved" and its analogs, favored by science journalists and other "outsiders," from which it seems to follow that properly objective (i.e., respect-worthy) knowledge will adhere to such a formal scientific standard. It will not bear the marks of its maker(s): it could be anyone's or everyone's and it will, in consequence, claim universal validity. Commonly, even in white Western-Northern affluent societies, nonscientists have little choice but to consider, debate, and endeavor to evaluate such reported testimonial findings "from outside," by way of informing and animating their/our beliefs and conduct. Quotidian repositories of knowledge rarely expose their sources and allegiances: the power of the "narrators" is undeniable, yet usually invisible, and not readily contestable.

As Oreskes and Conway read them, the articles in the resource book create an "impression of science rife with exaggeration, mismanagement, and fraud."[18] In consequence, it is no wonder that the judging

---

that the 'best solutions come from people making their own choices in a free marketplace, rather than government intervention.'" *Merchants of Doubt*, p. 147.
17. Oreskes and Conway, *Merchants of Doubt*, p. 6.
18. Oreskes and Conway, *Merchants of Doubt*, p. 146.

process for nonscientists (and likewise for many scientists) is confusingly complicated, reliant as it is on multiple layers of conflicting, presumably neutral testimony en route to determining whom or what to trust and how. Readers may discern a dispassionate if not disinterested approach in the EPA scientists' research; whereas, by contrast, in the readings the authors cite, *Bad Science* seems to rely on cautionary quips and throwaway lines about betrayals of public trust and costly policy decisions, infused before the fact with the very political beliefs the deniers are seeking to establish.[19] Such impressions scarcely suffice as the bases for informed judgment. Oreskes and Conway acknowledge that *Bad Science* does quote "experts" while noting that many of these experts were paid consultants to industry, an involvement that is rarely, and often only obliquely, announced in their assessments. In the main, however, the text adopts "a more sophisticated strategy: reminding readers of the fallibility of science."[20] My purpose is emphatically not to suggest that Oreskes and Conway have provided the only resources available to an interested and intelligent nonscientific public, but to present their work as part of an emerging state-of-the-art twenty-first-century Anglo-American area of inquiry of a credentialed epistemic community as a catalyst for ongoing research and debate.

A similar strategy to those just cited is also apparent in contributions to the aptly named, likewise path-breaking, and justly respected text *Agnotology: The Making and Unmaking of Ignorance*.[21] The book's purpose, albeit with variations from essay to essay, is to expose and unmask corporate and other widespread Western-Northern endeavors to promote and uphold public ignorance of the effects of anti-ecological thought and action. Such attempts trade on a margin of fallibility, of uncertainty, to sustain a level of public-social doubt that would caution *against* initiating or endorsing regulatory measures. The tobacco industry, pharmaceutical companies, and climate change deniers figure prominently in these analyses. Indeed, as David Michaels contends, "Uncertainty is

---

19. One "MESSAGE" reads: "Proposals that seek to improve indoor air quality by singling out tobacco smoke only enable bad science to become a poor excuse for enacting new laws and jeopardizing individual liberties" (Oreskes and Conway, *Merchants of Doubt*, pp. 144–45).

20. Oreskes and Conway, *Merchants of Doubt*, p. 146.

21. Robert N. Proctor and Londa Schiebinger, eds., *Agnotology: The Making and Unmaking of Ignorance* (Stanford: Stanford University Press, 2008).

manufactured. Its purpose is always the same: shielding corporate interests from the inconvenience and economic consequences of public health protections."[22] Either way, and this becomes the crucial point, the judicious response cannot plausibly be to conclude that because science is frequently marked by degrees of uncertainty, because it is fallible, there are no good reasons to heed such warnings. Ongoing extended and multivocal inquiry, public deliberation, debate have to remain high on the social-political-ecological agenda, together with reformed evaluations of the processes of judgement that inform and commonly shape the relevant inquiries and the credibility criteria, in self-satisfied societies. The point is not that all action informed by such affirmations must be held in abeyance until absolute certainty is achieved, but rather that holding spaces open for ongoing precautionary deliberations is vital to a collectively well-functioning, epistemically responsible society.

For nonscientists and for members of a larger population who, even despite their/our best efforts, can rarely claim a level of scientific literacy sufficient to qualify them/us to evaluate contradictions between well-informed climate skeptical positions, and ideologically motivated cherry-picking of evidence to serve climate skeptical ends, the situation can be frustratingly puzzling. One obvious and powerful, if rarely adopted, approach, in addition to on-going individual and collective investigation, is to promote and animate public conversations, consultations, town hall meetings, and collaborative debates toward developing effective *advocacy* practices—and places—that could facilitate navigating a world that is increasingly incomprehensible to nonspecialists, nonexperts. Such a proposal will meet with resistance from people for whom advocacy is, by definition, a tainted practice. Michaels, for example, notes: "Opinions submitted to regulatory agencies by corporate scientists . . . must be taken as advocacy, primarily, not as science."[23] Yet here, I suggest, he is presupposing a false dichotomy between science as knowledge conveying, and advocacy as mere propaganda, whereas the problem at issue is about advocating responsibly or otherwise, and where *epistemic* responsibility—a responsibility to know well, collectively and singly—and humility, as integral to policies and practices of honourable advocacy, are installed as *sine qua non* ingredients. These ingredients figure prominently in the

---

22. David Michaels, *Doubt is Their Product: How Industry's Assault on Science Threatens Your Health* (New York: Oxford University Press, 2008), p. 96.
23. Michaels, *Doubt Is Their Product*, p. 102.

self-definitions and in the practices of responsible inquirers, in both their individual and collaborative epistemic agency. Determining where and how, reasonably, to trust testimonial evidence that participants may understand only partially, is the hardest task: advocacy can contribute well or badly to the cultivation of understanding and awareness, but it cannot be presumed villainous or virtuous before the fact—an insight I return to in the next chapter. Nor, at its best, can it be instantaneous. Clearly, at the very least, communal, collaborative deliberation has to be an open, available option: epistemic individualism—the view that even the uninitiated can know by their/our solitary, independent endeavours—comes up sharply against its own limitations.

Epistemic Identity and Situation

Against a backdrop of tacit yet persistent Western-Northern convictions that the epistemological questions infusing these debates are generically *human*, to which neither sex, nor gender, nor disability, nor race, nor class, nor any other specificities of human "identity" pertain, I turn to a second, equivalently vital concern: about subjectivity and agency. Such matters, I propose, are constitutive—if too often silently so—of practices of giving and receiving testimony, in an expanded sense where testimony includes (but is not limited to) exposure to hitherto unconsidered points of view, receiving knowledge/information from other people, whether in conversations/debates; in media reports and enactments, or in reading, films, television, and myriad other sources. Sex-gender issues may seem to be irrelevant to these debates: the *Merchants of Doubt* text tacitly presents them thus and is not unusual or straightforwardly at fault in so doing. Its political motivations are laudable for their contributions to practices of thinking ecologically and to opening paths of inquiry that shift the directions of sedimented practices from isolated top-down analyses to incorporate lateral considerations, issues that stretch across human and situational similarities and differences. Still, there are highly valuable, activism-informing insights to be gained from taking subjectivity into account in all of these circumstances, especially in light of the tenacity of residual, unreconstructed public images of "science" as impersonally objective and "the scientist" as an infinitely replicable, neutral placeholder. The "subject"—hence the science-making subject(s) and the subject seeking to defend *his* freedom—makes few explicit appearances in the *Merchants of Doubt* text, even though scientists on both "sides" of the issues are mentioned by name and credentialed accordingly. Yet recurrent worries about freedom that run through this tale of the perpetuation of doubt expose the effects of a tacit conception of human subjectivity which merits closer investigation. To repeat, it is too rarely observed that, at least in white Western-Northern societies, the freedoms abstractly invoked and zealously defended by climate change deniers are emphatically *not* equally distributed across a sex/gender—or any other—social-political order. That political inequality which infects the freedom that skeptics defend makes the "liquidation of the subject"[24] from much of the text still more striking. For all its impressiveness, Oreskes and Conway's inquiry is limited for ecological-sexuate-gendered purposes by the almost-but-not-quite erasure of the subjectivities of those on whom challenge and change depend. The "human" generically conceived is only obliquely present in this text, stripped of its subjectivity and, *a fortiori*, of its gendered and other constitutive specificities. In short, the question "Whose *freedom* are we talking about?" moves into focus only in projects that aim to realize the transformative-disruptive potential of Oreskes and Conway's investigations. Recall Val Plumwood's long-ago reminder (following Luce Irigaray) that women "provide the environment and conditions against which male 'achievement' takes place, but what they do is not itself counted as achievement."[25]

Because the investigations Oreskes and Conway report take place behind a mask of impersonal epistemic replicability, appeals to specifically sexuate, situated, ecologically imagined subjects and practices have to negotiate through fixed assumptions about human sameness and the enhanced epistemic reliability of dislocated research. Such appeals are not easily inserted into an individualist-masculinist yet putatively gender-neutral frame of reference, nor is the issue just about individual reform. It is likewise about communal-collaborative-collective social change that requires destabilizing an entrenched imaginary of individualistic mastery and control, all the way down. Thus, revisiting a simple but persistent sexed-gendered connection, such disruptions will

---

24. The phrase is from Linda Martín Alcoff (following Horkheimer), "Epistemologies of Ignorance: Three Types," in Shannon Sullivan and Nancy Tuana, eds., *Race and Epistemologies of Ignorance* (Albany: SUNY Press, 2007), pp. 53–54.

25. Val Plumwood, *Feminism and the Mastery of Nature* (New York: Routledge, 1993), p. 22.

begin by recognizing that "mastery" has to be understood in light of the historically presumptive *maleness* of the masters who, when/if they make space for women on the turf they have jealously staked out as theirs, still in the twenty-first century, do so on their own terms, relegating such women to the margins: the gravel. (Let us remember that one striking difference between masculine and feminine approaches to knowledge, according to Margaret Whitford is that "to speak or write like a man is to assert mastery, to claim truth, objectivity, knowledge, whereas to speak like a woman is to refuse mastery, to allow meaning to be elusive or shifting, not to be in control, or in possession of truth or knowledge. So to be assertive to make claims, to be 'dogmatic,' which means to have a thesis, a meaning, a political position, is to take up a 'male' stance, whatever one's sex."[26])

In mainstream Anglophone epistemology, tacit conceptions of "freedom" and "achievement" are so closely aligned as to be co-constitutive. Yet the point, too tersely put, is that expectations of and assumptions about freedom and constraint take for granted and are informed by a cluster of habitual enactments of subjectivity that are neither neutral, laudable, nor universally realizable; nor are they, before the fact, open to being considered politically innocent. Even when such enactments remain tacit, they are substantively integral to producing and sustaining a wide range of social-epistemic practices, to such an extent that they can neither be left unaddressed nor remain *hors de question*.

Feminists and other marginalized Others from an unmarked white affluent masculine norm have long—and rightly—insisted that all freedom is someone's freedom, just as all knowing is someone's knowing, where "someone" could be singular or plural—or singular because plural. Either way, it matters who this subject is (or these subjects are) for understanding the production, circulation, uptake, evaluation, and enactment of knowledge claims. Feminist epistemologists have been engaged since the early 1980s in showing, to varying degrees, how the "sex of the knower" plays a constitutive part in such processes:[27] it is within the spaces their/

---

26. Margaret Whitford, *Luce Irigaray: Philosophy in the Feminine* (London: Routledge, 1991), p. 50.

27. Space does not permit an elaboration of these claims, but feminist epistemologists have, since the 1980s, advanced sophisticated, nuanced arguments in support of this assertion. For state-of-the-art analyses, see the essays in Heidi Grasswick, ed., *Feminist Epistemology and Philosophy of Science: Power in Knowledge* (New York: Springer, 2011).

our analyses make available that I will reread Rachel Carson's life and work later in this chapter. Evidently and without contest, the freedom the merchants of doubt are determined to protect is principally available—still now in 2020—to propertied, heterosexual, adult white men, in affluent Western-Northern societies. Hence "taking subjectivity into account"[28] in thinking about how ecological—and myriad other matters—are known requires taking sexed/ gendered/ raced/ classed/situated subjectivities into account, critically and genealogically, in the diverse specificities of their time and place. The purpose is in no way to reenact a superficial identity politics in epistemology or beyond. It is to excavate and evaluate subterranean forces that have shaped how the doing, knowing, being, dwelling at issue are enacted, there, and to analyse—genealogically—the implications of "situatedness" for knowledge-making projects. This undertaking cannot be a one-off instantaneous or "individual" endeavor: it requires time, study, investigation, collaboration, conversation, and a significant measure of *epistemic humility* to guard against any "rush to judgment."

A further insight emerges from the impersonal nature of Oreskes and Conway's analysis. As Plumwood observes and I too have noted, integral to the social imaginary that sustains mainstream Anglo-American philosophy is an assumption that "the human" is implicitly masculine, not just conceptually but in its multiple and diverse effects, "while the feminine is seen as a derivation from it."[29] Likewise, the contestable "autonomy" ideal underlying the pleas for freedom Oreskes and Conway narrate—but do not engage—has, throughout the history of modern white Western-Northern liberal political thought, derived from and celebrated possibilities afforded by and to affluent white heterosexual adult male/ masculine lives.[30] Freedom to find himself "at home" on the turf is an emblematic affirmation of the maleness of this ideal, available only to

---

For "the sex of the knower," see my "Is the Sex of the Knower Epistemologically Significant?," *Metaphilosophy* 12, 3/4 (July/October 1981), pp. 267–76.

28. The reference is to my essay "Taking Subjectivity into Account," in my *Rhetorical Spaces: Essays on Gendered Locations* (New York: Routledge, 1995), pp. 23–57.

29. Plumwood, *Feminism and the Mastery of Nature*, p. 23.

30. With the possible exception of men made frail by age. For a pertinent critique of autonomy, see my "The Perversion of Autonomy and the Subjection of Women: Discourses of Social Advocacy at Century's End," in Catriona Mackenzie and Natalie Stoljar, eds., *Relational Autonomy: Feminist Perspectives on Autonomy, Agency, and the Social Self* (New York: Oxford University Press, 2000), pp. 181–212.

those who need not notice the mundane tedium of *allgemeine Alltäglichkeit* with its routine disdain for the repetitious reproductive labor of quotidian domesticity.

Even after decades of feminist theory and practice, these vital considerations remain urgently underacknowledged; their effects too rarely taken into account in putatively universal analyses. They need, again now, to be critically revisited, especially when appeals to a generic *freedom* are invoked to excuse or condone such ecologically and other socially-politically destructive practices. As to how this revisiting could evolve and what its effects might be, it will have, in large part, to be an engaged, collaborative, deconstructive, and genealogical project of exposing the multiple, diverse sources and power-infused social-political effects of tacit yet entrenched assumptions about whose knowledge matters and can claim acknowledgment, what kinds of knowing achieve credibility—even hegemonic status—and why; which knowledge claims rightly count as exemplary—as standard-setting—for epistemology and beyond.

Chris Cuomo contributes provocatively to thinking well about such matters. She observes that "sexism, colonialism, racism, and harms to nature are practically, causally, and conceptually linked, and specific male-dominant cultures have been and continue to be the leading drivers of modern environmental destruction and degrading social violence."[31] Her focus on the gendered aspects of global injustice contributes significantly to ongoing deliberations from this relatively new gender-cognizant standpoint, whose very newness—too—invites critical-constructive engagement. A gender-cognizant standpoint is similarly if variously pursued in the *Agnotology* volume (Proctor and Schiebinger 2008).[32] Gendered considerations may seem not to matter for such self-contained, detached, and seemingly neutral examples as the cups on tables or cats on mats beloved of classical empiricists, or for the fake barn facades of some contributors to newer empiricisms. But they do matter for the modalities of social knowledge that inform or thwart ecological thinking and practice in ways that reinforce entrenched injustices and power structures;

---

31. Chris J. Cuomo, "Sexual Politics in Environmental Ethics: Impacts, Causes, Alternatives," in Stephen M. Gardner & Allen Thompson, eds., *The Oxford Handbook of Environmental Ethics*, Oxford University Press, 2017, p. 289.

32. See Robert Proctor and Londa Schiebinger, eds., *Agnotology*. Especially relevant in this regard are the articles by Charles Mills, Londa Schiebinger, Nancy Tuana, and Alison Wylie.

and they do matter for projects of sustained critical engagement with the politics and epistemology of testimony. Often and without contest, they take for granted an easy, uncontested availability of the items and actions they present as paradigmatic. Hence, again, the pertinence of Cuomo's contribution.

From an ecologically informed stance, Mick Smith, in *An Ethics of Place*, offers an analysis of this situation that confirms the view outlined here. I quote him at length:

> As modernity's offspring we . . . tend to understand our own identities and social relations . . . [through an] atomistic ideal/ideology of autonomous, bounded, individuality . . . [as] concrete and isolable individuals each on their own disparate trajectories, each with *particular* identities derived from . . . certain essential, quantifiable, and indefeasible properties. We are born under the sign(s) of one-dimensional *man*; *Homo economicus*, that self-contained and self-serving caricature of modern humanity, a parodic recapitulation of the instrumental order of capitalism and phallocracy. . . . Only "man" is intrinsically valuable. Women and nature are made subject to reason's cold calculations, their reality recognized only insofar as they become hard currency to be valued and traded according to their use.[33]

Recalling the passage from Virginia Woolf with which I began: clearly, by these standards, women have "no real being," and although such thoughts will be old news to feminist theorists and activists (and to ecological thinkers), they persist.

My comments from the end of chapter 1 merit reiteration here. Taking subjectivity seriously, and thereby disrupting and resisting masculinist social imaginaries of mastery, has profound ethical-political and epistemological implications. A responsible response to insights that emerge from gender-cognizant and other subjective standpoints will involve rethinking and renewing how subjectivity is conceived and enacted. It will involve reimagining ourselves—we who live in the affluent West—not merely in revisionary ways, but in ways that are sufficiently radical to disturb the

---

33. Mick Smith, *An Ethics of Place: Radical Ecology, Postmodernity, and Social Theory* (Albany: SUNY Press, 2001), p. 171 (italics original).

social-political-epistemic imaginary that grounds our lives and thoughts. Nor, I suggest, is the issue quite as Smith puts it in urging an "*alternative* conception of subjectivity"[34] for the language of "alternatives" is misleading in its implication that all (or at least many) ways are up for grabs. We can opt for one or another, interchangeably and intermittently, as we would select from a smorgasbord of edibles. This caveat notwithstanding, an activism informed by Smith's recommendations could work singly and collectively to unmask, discredit, and displace that "caricature of modern humanity," exposing it for the dangerous illusion it is through on-going, piece-by-piece deconstructions of its contributions to producing the ecologically unsustainable conditions that prevail in the affluent white Western/Northern world, to undermine its ontological and—in consequence—its epistemic credibility. Feminist theory and practice, singly and in their alliances with other postcolonial movements of the late twentieth and early twenty-first centuries, are engaged in just such displacement-revisionary projects. At their best, they/we are aware that these projects, if they are to succeed, have to proceed hermeneutically, understanding and contesting who *we* are just as fundamentally as they contest the social-economic structures we make and are made by. They/we have not completed this project: it may be unrealizable in its anticipated totality, but it cannot be ignored or abandoned.

The issue is ontological, and practical, ethical, political, and epistemological. Sexuate practices are peculiarly well equipped to animate the engagement such a radical re-visioning requires because they offer a particular line of vision, a way of seeing that, I suggest, evinces certain affinities to W. E. B. Du Bois's "phenomenological concept of double consciousness,"[35] which Du Bois himself characterized as "a sense of always looking at one's self through the eyes of others, of measuring one's soul by the tape of a world that looks on in amused contempt and pity."[36] This imposed incongruity with/in oneself was for Du Bois a product of oppression, and a source of ongoing agony. But Ernest Allen Jr. proposes revisiting the implications of "double consciousness" in ways that suggest

---

34. Smith, *An Ethics of Place*, p. 173 (italics added).

35. Cf. Lewis R. Gordon, *Existentia Africana: Understanding Africana Existential Thought* (New York: Routledge, 2000), p. 38.

36. Quoted from Ernest Allen Jr., "On the Reading of Riddles: Rethinking Du Boisian 'Double Consciousness,'" in Lewis R. Gordon, ed., *Existence in Black: An Anthology of Black Existential Philosophy* (New York: Routledge, 1997), p. 51.

something of the ontological deconstruction feminist women have also to perform because we too, albeit with radically different implications, have had to judge ourselves through the eyes of others. In a recommendation that echoes some of the tenets of standpoint epistemology, Allen notes that "rather than *celebrating* an authentic 'dual consciousness' as a tool for achieving enriched cultural or political syntheses, or as a platform for generating multiple levels of understanding—in other words, as a potential solution in whole or in part—Du Bois has treated the question of "twoness" chiefly as a (real or imaginary) problem, even as he affirmed the desirability of preserving certain of its (unspecified) forms."[37] At the risk of helping ourselves to concepts forged in forms of oppression that cannot have been ours, and of performing epistemic violence in so doing, I suggest there is something to be learned by analogy for feminists and other Others now (in the early decades of the twenty-first century), about sexuate being, from Allen's reading of the ontological and epistemological power in Du Bois's thought in both its negative and its positive connotations. In fact, there is a notable precedent to this proposal in Maria Lugones's landmark and poetically rich analysis of "world-travelling,"[38] where she urges feminists and Others to engage in careful, imaginative-creative practices of attempting to enter another's "world," to begin to understand, tentatively, if rarely conclusively, how it is to live in that world, those multiple worlds that enable her to *be*, to play—though not in any mocking or frivolous sense—with the enabling and constraining aspects of that "world." The implication is that the world traveler will not—should not—emerge unchanged from the journey. Something akin to these ideas is implicit in my thoughts about rethinking who we are.

Finally, as I observe earlier in this chapter and elsewhere, the example of Rachel Carson is impressively instructive for thinking about and working to understand connections among expertise, testimony, and subjectivity in the twentieth- and twenty-first century Western-Northern world. She lives and enacts ecological thinking as a way of inhabiting the world that confers content on the term, the idea, and the practice in ways consonant with Verena Conley's claims for ecological subjectivity as "relating consciousness of the self to that of being attached to

---

37. Allen Jr, "On the Reading of Riddles," p. 51.
38. Maria Lugones, "Playfulness, 'World'-Travelling, and Loving Perception," *Hypatia: A Journal of Feminist Philosophy*, 2, 2 (1987), pp. 3–19.

and separated from the world"[39] and where she endorses Ilya Prigogine's and Isabelle Stengers's image of human beings as "creatures immanent to their environment rather than fully autonomous conscious subjects," and Michel Serres's evocative observation that human beings "do not just dwell as individuals, they weigh on the earth." In such thoughts, critical attentiveness to ways of being contests unexamined ontological presumptions of human sameness, and of the liberal unified self.

Rachel Carson appears as an iconic figure in much of my writing and thinking about matters ecological, and appropriately so. As I observe earlier, her enduring emblematic status is confirmed and celebrated in the 2017 documentary film *Rachel Carson* directed by Michelle Ferrari. In her writings and practices, as I have shown in *Ecological Thinking*, Carson figures as a quintessential exemplar of *ecological* subjectivity.[40] For Conley, the urgent task for ecological thinking is to unmask "mass-produced subjectivity in societies of control with their consequences for natural and social ecology," a task as pertinent two and more decades later as it was when first she crafted it.[41] Such ways of thinking contest sedimented yet unexamined ontological presumptions about human sameness, freedom, and the autonomous liberal unified self. They are as normative as they are descriptive. Hence, it is also worth emphasizing that even though the natural world significantly constrains approaches and points of view that can achieve consensus within a going ecological imaginary in interpreting and understanding ambiguous data and managing uncertainty, Eric Biber is right to note that "[w]here reasonably possible, scientists tend to interpret their observations as consistent with whatever theory currently commands the most adherents, even if other interpretations are equally or even more plausible."[42] As theorists of ecological subjectivity increasingly

---

39. Verena Andermatt Conley, *Ecopolitics: The Environment in Poststructuralist Thought* (London: Routledge, 1997), p. 10.

40. If there are residual doubts about her enduring scientific-feminist significance, the 2017 film *The Power of One Voice: A 50 Year Perspective on the Life of Rachel Carson*, directed by Mark Dixon (US, 2014) tells an impressive, carefully situated historical-intellectual story-biography of Carson's scientific, wider epistemic, and personal life, showing her bravery in the face of the persistent—if veiled—sexism she encountered, among numerous obstacles in the then-new male-science-venerating world.

41. Conley, *Ecopolitics*, pp. 74–75.

42. Eric Biber, "Which Science? Whose Science? How Scientific Disciplines Can Shape Environmental Law," *The University of Chicago Law Review* 79, no. 2 (Spring 2012), p. 503.

affirm, such points of view are as often animated by dominant (if tacit) values that shape and sustain the discipline as they are by situational factors. Those values, too, have to be held open to analysis and critique, even if the fact of their influence seems not in itself to be reprehensible. Rachel Carson, then, is not just the ecological subject but also, in ways pertinent to the claims I advance here (borrowing the term from Luce Irigaray), she is the *sexuate* subject albeit, in the climate of her times, often silently, tacitly so, and with positive and negative implications.[43] Evidently, in herself, Carson regarded neither her way of life nor her scientific practice as shaped or otherwise influenced by her female/feminine being. In fact, no conceptual-discursive space was available for thinking and/or addressing such issues during her lifetime, nor for some years after her death.[44] Fortunately, in the discourse of feminist and postcolonial epistemology and philosophy of science now, her scientific practice, for all the regularity of its approach and competence, hovers on the edge of attesting to a female/feminist standpoint and is often discredited accordingly even where it merits veneration and acclaim. Some of her best, most ecologically sophisticated work sits just here, in a not-yet-realized sex-gender-specific frame and style of reasoning.[45] It is she who exemplifies the power and the perils of an informed advocacy that can be a crucial piece of sound ecological practice. (Her scientific-investigative style contrasts sharply with the *Bad Science: A Resource Book* agenda, in ways that have garnered both praise and ridicule.)

For a time, Carson achieved impressive scientific and public legitimacy and acclaim, if only uneasily and precariously. Ironically, she died

---

43. Whitford, *Luce Irigaray*, p. 50. Specifically, "to speak or write like a man is to assert mastery, to claim truth, objectivity, knowledge, whereas to speak like a woman is to refuse mastery, to allow meaning to be elusive or shifting, not to be in control, or in possession of truth or knowledge. So to be assertive, to make claims, to be 'dogmatic,' which means to have a thesis, a meaning, a political position, is to take up a 'male' stance, whatever one's sex."

44. The allusion is to Michel Foucault, "The Discourse on Language," in his *The Archaeology of Knowledge and the Discourse on Language*, trans. A. M. Sheridan Smith (New York: Pantheon Books, 1972), where he observes "Mendel spoke the truth, but he was not *dans le vrai* (within the true) of contemporary biological discourse" (p. 224).

45. "Styles of reasoning" is Ian Hacking's phrase in "Language, Truth and Reason," in Martin Hollis and Steven Lukes, eds., *Rationality and Relativism* (Cambridge, MA: MIT Press, 1982), pp. 48–66.

from breast cancer, in other words, from one of the ecologically most devastating effects of profligate uses of DDT. Now in the twenty-first century, she is both celebrated anew and cast out as the sacrificial subject, vilified and castigated as a murderer. It is also noteworthy that she is the only epistemic subject who claims a fully narrated place in the *Merchants of Doubt* text, although it should be observed that she is not the only scientist to be identified by name by Oreskes and Conway, nor is she the only one whose qualifications and training are cited to validate her/his testimony. Climate change defenders and deniers, too, are identified with and by their specific credentials.

What, then, can be the point of singling Carson out, yet again, for extended discussion? Although, as I note, most of the scientists whose work is discussed in the *Merchants of Doubt* text are named, and their credentials and allegiances are detailed (even some of those they attempt to conceal), Carson is the only one who *in herself*—contrasted with *for her work*—becomes a focus of attention. The legitimacy of other scientists' place-holder status in claiming "truth, objectivity, knowledge" is neutrally assumed by Oreskes and Conway, despite bitter disagreements surrounding the content and implications of these scientists' hypotheses and findings. In the gendered/racialized twenty-first-century politics of knowledge, this contrast is significant. Presumably because she is white, no comment on Carson's race is called for: whiteness then, and still frequently now, counts as the default norm, at least in much Anglo-American philosophical and other scholarly writings.[46] Nonetheless, it can scarcely be a coincidence that the only female environmental/ecological scientist discussed at length in this book is apparently judged—albeit on the whole quite favorably—as much by her "person" as by her work. Analyses both critical and commendatory seem unable to avoid referring to her sex or to refrain, implicitly or explicitly, from factoring it in to their evaluations, sometimes favorably, and when negatively, only obliquely. Few commentators fail to note that Carson was a single (i.e., unmarried) woman—a "spinster."

This observation is not intended to charge Oreskes and Conway with offering a sexist reading; quite the contrary. Yet it is meant to note, if cautiously, that Carson herself is more plainly visible in her work than are most of the scientists the authors name and discuss, and hence to suggest that her sex/gender in this respect seems to be perti-

---

46. See Linda Martín Alcoff, *The Future of Whiteness* (Cambridge, UK: Polity, 2015).

nent. In the scientific circles where she was active, it could as readily have been presented as the focus of laudatory as of condemnatory or silently dismissive discussion. Hence, the political effects of attending so closely to Carson and to the nature of her working practice are positive in several respects: in attending to the scientific exceptionality—in two senses—of Carson herself and of her achievements against near-impossible quotidian and professional obstacles; in commending her often-solitary findings despite her having virtually none of the institutional and public support a male inquirer could routinely expect at that time; in persevering against charges only recently recognized as sexist and *ad feminam*. But these observations tell only part of a larger story. She is without doubt an ecological and feminist hero.

Still, a peculiar vulnerability often attaches to the situation of a woman in science, now almost three decades into the twenty-first century and dramatically more so for a woman of Carson's time, especially one whose financial and family care-giving situations prevented her enrolling in doctoral studies. Such *personal* vulnerability is well known, and it is not difficult to imagine (despite the difficulties that attach to eradicating it.) Less well known, and still more significant for purposes of this discussion, is the *epistemological* vulnerability that attends and/or grows out of it. Sex-specific charges against Carson's work and thought were routinely covered over in her time, but they were there, if sometimes veiled as they are again in the twenty-first century, in critiques of female and other nonwhite, nonmale intellectual endeavors too numerous to recount. Accusations of hysteria proliferate. Where female-engaged or -directed projects are at issue, such accusations are frequently enlisted to "explain away" their findings, especially when such findings fly in the face of widespread sedimented assumptions about "real science." For Oreskes and Conway, in whose analyses such charges are multiply detailed, the accusations have as much to do with Carson's sex as with epistemically irresponsible accusations that she was doing bad science—yet where, on closer examination, it becomes apparent that the irresponsibility has less to do with the substance of her research and more with the attackers' and discreditors' failures to base their charges on adequate investigations of the minute yet on-going, evolving complexities of the effects—for example, of DDT—which are neither as uniform nor as static as the "evolution" of rocks might be.

Still more germane to the dismal quality of the critics' assessments of her research than their having made a scapegoat of Carson herself is their failure to take into account such effects as "the well-documented and easily found (but extremely inconvenient) fact that the most important reason that DDT failed to eliminate malaria was because insects *evolved* . . . a truth that those with blind faith in free markets and blind trust in technology simply refuse to see."[47] Without doubt, this rush to judgment is ethically and epistemologically reprehensible. Oreskes and Conway observe that to know whether the deleterious effects of DDT for female reproductive health were significant enough to support continuing to ban it, it was vital to engage in longitudinal studies, for example, of women who had been exposed early in life, "when environmental exposures where high."[48] One such study, conducted in 2000–01 on women then in their 50s and 60s who had been exposed to DDT as children or teenagers, "showed a fivefold increase in breast cancer risk among women with high levels of serum DDT or its metabolites."[49] If such findings can continue to emerge long after the effects of DDT have weakened in malarial districts, then they contribute after the fact to vindicating Carson's longitudinal approach, even as they urge further departures from the instantaneous spectator epistemology of standard empiricism and toward "horizontal" as contrasted with vertical, top-down analyses: to taking a longer view across populations, terrains, and time-frames, and engaging in extended hermeneutic analyses, when the subject matter is appropriate.

By contrast, characteristic of Carson's scientific "style" is her wariness of too-ready translation from one domain or species to another of insensitive, often too-swift classification practices that may judge too quickly, thereby failing to notice or to count apparently minor differences, aberrations from a norm. Although in her scientific practice she undoubtedly understands the allure of mastery and control, her ways of achieving it often require slowly, painstakingly following up on narrow and/or precise local hypotheses that differ significantly for each of the species she studies—as is apparent in her research on the Japanese beetle, the gypsy moth, Dutch elm disease—as the hypotheses differ also for investigating the short- or long-term implications of chemicals for human health. Catching a central contrast between an overarching ethos of mastery and an ecological ethos, Carson deplores a stubborn corporate

---

47. Oreskes and Conway, *Merchants of Doubt*, p. 236.
48. Oreskes and Conway, *Merchants of Doubt*, p. 229.
49. Oreskes and Conway, *Merchants of Doubt*, p. 239.

and more widely dispersed resistance to taking a longer view—to waiting "an extra season or two"[50]—when a quick (chemical) fix is ready to hand. Ecologically, thinking as she does requires factoring time, place, and history into responsible scientific investigations. As I understand it, this is an important epistemological requirement. It bears directly on Carson's thinking about causality, where she also departs from standard epistemological assumptions and challenges them in so doing. And it points toward a need for closer engagement with, and an elasticity in working with, matters of *time* in philosophy of science, ecological thought, and climate change epistemology. This is a delicate matter to endorse unequivocally in climate change discussions for despite the claim's initial plausibility, climate change activists insist, often appropriately, upon the urgency of immediate action, arguing that whatever the uncertainty, action is required, *now*, based on the best available predictions. "We" may not have even a season to wait. Carson, in the 1950s and 1960s, worked from the (perhaps tacit) conviction that causal connections may not be immediately apparent to people who are neither informed—nor prepared—to look ecologically. Thus, it is an easy matter for investigators to discount causal claims that extend temporally and geographically away from a specific chemical application, claims that require imagination, conjecture, patience, and time for confirmation or falsification. Her ecological approach vindicates such a longer view. Yet the hypotheses she works from can, in numerous situations, guide inquiry whose empirical generalizations stand up well against the quick and dirty solutions proposed by the chemical industry and its champions.

Whether Carson's diverse achievements are enabled or enhanced by her sex, whether they can be attributed—positively—to her being a woman, is a different, more complex question. Many aspects of her scientific practice and her writing style, some of which I have mentioned, contrast sharply with the orthodoxy of received "scientific" practice. But attributing gendered significance to them is not a straightforward matter, even though her work displays affinities with explicitly feminist twenty-first-century epistemological inquiry. And yet as the 2017 film I have mentioned confirms, if obliquely, various aspects of her "being" so to speak, contributed to her struggles for recognition: not least among them the fact that—albeit primarily for economic and other circumstantial

---

50. Code, *Ecological Thinking*, p. 42.

reasons—she was a woman who had no PhD. More to the point and again germane to some of the after-the-fact gender-specific claims is the extent to which Carson was criticized for her *styles* of research and reasoning rather than for its substance and the extent to which some negative consequences of her research were laid at her feet as explicitly *ad feminam* charges. Carson, as Oreskes and Conway note, "documented at great length both the *anecdotal* and systematic scientific evidence that DDT and other pesticides were doing great harm."[51] Ironically, her respect for the "anecdotal," which is integral to her practice, and is often effective, has contributed to the fluctuating respect and easy vilification in the condemnatory rhetoric of her detractors. It connects with the rhetorical, conceptual architecture of the epistemic-scientific world where Carson worked, structured as that world was (and still often is) by hierarchical divisions between fact and anecdote, truth and narrative, reason and feeling, of which the first item in each pair claims greater public and professional credibility, authority, and reliability than the second. The division locates anecdote, narrative, and feeling on the negative, subjective, feminized side where meaning can be "elusive or shifting," while for epistemologies of mastery it is in *facts* alone, dispassionately discovered, that truth is to be found. In her respect for down-on-the-ground experiential reports commonly dismissed as merely *anecdotal*, Carson's epistemic practice unsettles these distinctions—and earns her the label "subversive" in so doing.[52] Criticisms of Carson's respect for "anecdotes" are reflective of larger issues about testimony, and about how credibility plays out in the politics of knowledge. Credibility has, again, been tacitly coded "masculine" across a wide range of situations and practices. Hence, thought styles, styles of reasoning tainted with female/feminine associations can claim a place within "the credible," "the rational" only by conforming to formal, putatively disinterested dictates. Rereading Carson's work in the early twenty-first century, and evaluating its struggles to achieve uptake and acknowledgment when the epistemology and politics of testimony are beginning to animate vibrant developments in social epistemology, suggests that her reliance on what was dismissively labeled "anecdote" might now more readily have assured

---

51. Oreskes and Conway, *Merchants of Doubt*, p. 219.
52. Mark Hamilton Lytle, *The Gentle Subversive: Rachel Carson, Silent Spring, and the Rise of the Environmental Movement* (New York: Oxford University Press, 2007).

her a "respectable" place in new approaches to public knowledge and ecological investigation.[53]

Even though, in Rachel Carson's lifetime, such discrediting as I have mentioned was less often conveyed in sexed/gendered terms than it subsequently has been, the flavor was unmistakable. Now, it is less carefully masked. So, for example, Michael Smith, in an article tellingly titled "Silence, Miss Carson" (a title borrowed from an "unbalanced" review of *Silent Spring*) refers to a "prevailing attitude" for which she was "an uninformed woman who was speaking of that which she knew not. Worse, she was speaking in a man's world, in the inner sanctum of masculine science in which, like the sanctuary of a strict Calvinist sect, female silence was expected."[54] Nor can it be merely a coincidence that the most damning website devoted to discrediting Carson, which comes from a 2009 Competitive Enterprise Institute project, is called *rachelwaswrong.org*.[55] Apart from the contents of the items on the site, the chastizing "bad little girl" tone conveyed in naming it thus is egregiously demeaning: a woman too insignificant to be referred to by her full adult name has ventured too far onto epistemic territory that should not be hers, and is sternly, patriarchially reprimanded. This is the woman who in her time and still frequently in ours is dismissed as hysterical, or scare-mongering: condemned for putatively negative affect as having influenced her work. In short, to understand Carson

---

53. Code, "Particularity, Epistemic Responsibility, and the Ecological Imaginary."

54. Michael B. Smith, " 'Silence, Miss Carson!' Science, Gender, and the Reception of *Silent Spring*," *Feminist Studies* 27, no. 3 (Autumn, 2001), p. 736.

55. See http://rachelwaswrong.org/ For a fuller discussion of Carson's scientific practice see Code, *Ecological Thinking*, especially chapter 2. Oreskes and Conway note: "The Internet is flooded with the assertion that Carson was a mass murderer, worse than Hitler. Carson killed more people than the Nazis. She had blood on her hands, posthumously. Why? Because *Silent Spring* led to the banning of DDT, without which millions of Africans died of malaria" (*Merchants of Doubt*, p. 219). Noting "her legacy has been characterized as 'Rachel Carson's Ecological Genocide,' " Steve Maguire cites the relevant article: "let there be no mistake: Rachel Carson and the worldwide environmentalist movement are responsible for perpetuating an ecological genocide that has claimed the lives of millions of young, poor, striving African men, women, and children, killed by preventable diseases." "Contested Icons: Rachel Carson and DDT" in *Rachel Carson: Legacy and Challenge*, ed. Lisa H. Sideris and Kathleen Dean Moore (Albany: SUNY Press, 2008), p. 194.

as "a human being" and as a woman struggling to be a scientist in an inhospitable environment is to understand something of the *dis*-ease, the *un*-ease of that position: she is both an exemplary ecological subject—in large part to her triumph—and, in the eyes of the "neutral" scientific establishment, an exemplar of a reprehensibly sexed-gendered being, if neither one always in her own eyes. For Carson, manufactured uncertainty undoubtedly attends the reception of her meticulous work: uncertainty manufactured and sustained by a scantily veiled, largely unspoken, but—paradoxically—nonetheless blatant sexism. Skepticism or uncertainty toward Carson's work is not inherently troubling. Wendy S. Parker argues plausibly for the *value* of uncertainty acknowledged, as a wisely preventative stance in dissuading investigators from any rush to judgment; toward making poor, hasty decisions.[56] The issue with the skepticism Carson faced is, in my view, that the uncertainty is inappropriate since it is at root, discriminatory.

To conclude this section, I draw these lines of thought together to confirm my reasons for pursuing them. First, the growing significance of testimony in social epistemology in the early twenty-first century creates spaces for epistemologists and moral-political theorists to engage philosophically with situations in the divisions of intellectual-epistemic labor in white Western societies where "we" are commonly reliant on other people to "know" for us and where often, by this very fact, we do not and perhaps cannot know well enough to judge whether they are epistemically responsible or to discern how to place our trust wisely. These thoughts engage some of the most challenging issues in present-day mass-media-conversant theories of knowledge. Thus, I am reading the Oreskes and Conway text together with analyses of Carson's scientific practice, as events in the epistemology of testimony.[57] The chains of analysis can be long and interwoven, and certainty may be elusive. But such is the consequence of breaking away from an individualistic picture of knowing, and moving toward a community or communal one, where

---

56. Wendy S. Parker, "Environmental Science: Empirical Claims in Environmental Ethics," in *The Oxford Handbook of Environmental Ethics*, ed. Stephen M. Gardiner and Allen Thompson (New York: Oxford University Press, 2017), pp. 27–39.

57. Representative texts in the testimony literature are C.A.J. Coady, *Testimony: A Philosophical Study* (New York: Clarendon Press, 1992); Jennifer Lackey and Ernest Sosa, *The Epistemology of Testimony* (New York: Oxford University Press, 2006); José Medina, *The Epistemology of Resistance* (Oxford: Oxford University Press, 2012).

neither the composition of the community nor readers' capacities to evaluate it well can be presupposed before the fact.

For the merchants of doubt, impersonal gender-neutral "facts" are marshaled to promote or defend a faceless, dislocated freedom. They can effectively be countered only in complex, textured tales where, epistemologically, the task is both phenomenological and hermeneutic, and—again—needs to be performed with a certain humility. It is also empirical, but it is more effectively empirical when it is, at the same time phenomenological and hermeneutical: a textured, interpretive inquiry where the epistemological and the social-political cannot plausibly be held apart. A doubled consciousness such as may have been Carson's but must also be "ours" will come insistently but differently to bear on each of the many subject matters, subjectivities, and issues that have to be investigated if sustainable human and other animate and inanimate futures are to be promoted, against the insistence of merchants of doubt who are determined to gainsay them in the name of a freedom that is destructive at the core, even of the subjects who champion it.[58] Whether these denials can continue in the face of ongoing IPCC investigations and reports remains to be seen; the deniers will have to work relentlessly to manufacture levels of uncertainty sufficient to counter the force of these increasingly damning findings. Still, as Will Hutton observes: "[I]t will be met by a barrage of criticism from the new 'sceptical' environmental movement . . . which, while conceding that global temperatures are rising, insists that there is still insufficient scientific proof to make alarmist predictions."[59]

The larger implication is that initiating sustainable being, knowing, and dwelling requires a radical contestation of mastery and its arch-masculinity and toward a greater respect for the limitations of our knowledge and the need—where the urgency of the situation allows it—to develop a tolerance of ambiguity both in Simone de Beauvoir's sense and in the sense captured by Margaret Whitford and explicitly feminized. The task for us now—where "us" is the most contestable of these terms—is

---

58. Some sections of this chapter are drawn, with minor changes, from my chapter by the same title, in Peg Rawes, ed., *Relational Architectural Ecologies: Architecture, Nature and Subjectivity* (London: Routledge, 2013), pp. 73–90. They are reproduced with permission of Peg Rawes and the editors.

59. Will Hutton, "Our planet needs us to fight for its survival," *The Guardian Weekly*, 27 Sept., 2013.

to tread a narrow line between a reversion to conventional difference stereotypes with their capacity to keep women down and a celebratory enactment of female/feminist sexuate values and possibilities that might perform some of the social-political reenergizing the new social movements of the 1960s initiated and that are still available now as more sophisticated, more nuanced resources, to cultivate new subjectivities.

Second, I have urged acknowledging the need for communal, critical discussions and analyses of taken-for-granted assumptions about subjectivity and agency that prevail in white Western-Northern epistemic communities. For critically examining the place and practices of these communities in constructing positive epistemic exemplars or the reverse, it is instructive—imperative—to examine, geneaologically, how and why they claim or fail to claim that status: by what warrant or what withholding of respect. Advocacy, as I will discuss it in chapter 3, has in many respects been read as a feminized practice. It meets with resistance when its combative, legalistic (masculinist) truth-denying reputation drowns out its emancipatory potential—as though people in general, so to speak, were too stupid to practice, judge, and evaluate it well. People will argue for—advocate for—what they care about and/or fear to lose: both the deniers and the convinced engage in such practices, and advocacy good of its kind requires epistemically responsible, on-going investigations to inform and evaluate it. It has to be kept open to justification and/or contestation at multiple levels, in open communities of inquiry. Yet the response need not, and indeed *should not* be to condemn advocacy *tout court*, but to be vigilant for, to engage critically with, and to deplore its blindly aggressive and/or wilfully ignorant instantiations while enlisting it and acknowledging its value for social epistemology (which, in effect, means *all* epistemology).

Epistemic responsibility demands that "we" educators need, perhaps above all else, to learn to—and to teach our students how to advocatedebate responsibly, knowledgeably, and *humbly*, paradoxical as this need may be—in the minutely informed and ethically/politically respectful way Oreskes and Conway investigate: to recognize how zealously the deniers seek to defend places and putative values that are, quite simply, unsustainable. These deniers together form a tidal wave of capitalist opposition from those who think they have too much to lose and cannot see how much they have to gain—yes even of freedom, intelligently thought through. The challenge is to allow space for sexuate practices to contribute to fostering a kind of cohabitation from which a respectful way of *dwelling* could emerge that, from their very sexuate being some women have—somehow—been able to animate. Reading Carson as a quintessential ecological subject, whose sexuate being seems both to be one of the assets that shapes her practice and (in her time) one of the most challenging to counter, returns us to sexual difference. It even, as Alison Stone suggests, raises the possibility of acknowledging sexuate being as in some sense real (admitting but without belabouring its socially constitutive/constructed makeup); of affirming it and seeing in it a certain power and a perpetually beleaguered promise. The point is not to effect the reversal in gendered values that some feminists of the 1980s, following Carol Gilligan, deplored—it is to draw on, affirm, and exploit its values in situations that are usually local because of the meticulously detailed work required on all fronts: on the level of research, often into the smallest minutiae (as Oreskes and Conway detail) while needing to discredit the detractors/the merchants of doubt. Hence, advocacy too, with its care for the specific and particular case is in some of its aspects a feminized practice and it too meets with curious levels of resistance where its combative (masculinist) truth-denying reputation constantly drowns out its emancipatory potential. Integral to the bad press advocacy encounters is an entrenched postpositivist philosophical division between epistemology and ethics-politics, which allows insufficient space for acknowledging the urgency of epistemic responsibility as an intellectual *and* ethical value and commitment. Such a division, I suggest, is no longer plausible in the twenty-first-century Western-Northern world, especially now that social epistemology, with its (cautious) hospitality to interpretive, hermeneutic practices is demonstrating its effectiveness. In order to meet these challenges, I turn, in the next chapter, to probe more deeply into pressing questions about the nature of advocacy and its place in responsible knowing.

Chapter 3

Care, Concern, and Advocacy
Is There a Place for Epistemic Responsibility?

For philosophers situated within the philosophical/analytic mainstream, care has commonly been represented as a "feminized" practice, and thence as incapable of enacting objectivity. It is undervalued accordingly. This chapter will endeavor to reclaim the epistemic potential of responsible, knowledgeable care, together with humility and advocacy, regarding all three—together and separately—as epistemologically-ethically-politically valuable commitments which are integral to responsible knowledge and action. In their everyday iterations, they claim no specifically gendered associations. Yet here, and to the contrary, I will propose that *advocacy*, which is often represented as a tainted practice (allegedly/even plausibly for *its* failure to maintain a distanced, disconnected, neutral objectivity), can—and commonly does—contribute well to practices of acquiring and promulgating knowledge, thoughtfully if critically, across a range of social, political, ecological situations. It signals a commitment to interactive, socially-politically cognizant approaches to knowing and doing. Moreover, I will propose that *humility* claims a central place in multiple epistemic practices that center around matters of care and concern for the world both distant and close to "home," and for its diverse inhabitants. I will engage with it—with humility—as a vital contributor to thinking and acting well for/about inherently social-political inhabitants of a multiply diverse world. In that world, humility at its best is central to wise practices of listening carefully, arguing respectfully, reserving judgment, keeping spaces open for ongoing deliberation and debate: for intelligent receptivity and critique.

In thus advocating humility as a central intellectual and practical virtue, I am thinking, negatively, of an absence of arrogance; and positively, of an openness, a readiness to consider, discuss, and acknowledge the limitations of "our own" (single and collective) knowing, not in a groveling Uriah Heap sense, but in a spirit of respect, and of deference to the limitations of all knowing; to the consequent necessity of knowing responsibly and well. These initial thoughts may appear to convey a too-precious caution, but their practice need signal neither a distant and pious austerity nor a coldly aloof detachment. Practices that honor such diverse demeanors are vital aspects of care and concern as these currently inform moral philosophy in the Western-Northern world, and beyond.

Specific questions about care, concern, and advocacy that inform the thinking central to this chapter are prompted by a November 6, 2011, report in the Toronto daily newspaper *The Globe and Mail*: the story, by Mark Hume, of a controversy surrounding the ethics and politics of knowledge as they had been enacted in a Canadian courtroom. Entitled "Famous Medical Ethics Lecturer's Credentials Challenged in Euthanasia Case," the report engages with the question whether Dr. Marcia Angell could legally qualify as an expert witness in a debate about physician-assisted suicide.[1] Dr. Angell's official, public-professional credentials speak strongly in favor of her credibility as a witness, as do less formal accounts of her professional practice. She had been senior lecturer in the Department of Social Medicine at Harvard Medical School, where (at the time of the report) she "currently gives monthly lectures on ethics to faculty." She had been executive editor of the highly respected *New England Journal of Medicine* from 1988 to 1999 and interim editor-in-chief from 1999 to 2000. Author of an acclaimed book, *The Truth About Drug Companies*,[2] she had garnered widespread professional and public respect for her principled refusals to publish pharmaceutical-industry-funded research in the *Journal* while she was serving on its editorial board. These background facts are noteworthy not just in themselves, but especially for their pertinence to the questions about advocacy, expertise, and trust I consider in this chapter and to much larger issues of establishing and maintaining epistemic credibility in situations where vulnerability has carefully to be taken into account,

---

1. Mark Hume, "Famous Medical Ethics Lecturer's Credentials Challenged in Euthanasia Case," *The Globe and Mail*, November 6, 2011 (updated May 8, 2018).

2. Marcia Angell, *The Truth about Drug Companies* (New York: Random House, 2005).

at all levels of the deliberations. These facts attest further to Dr. Angell's suitability—professional, ethical, personal—to serve as an expert witness in a contentious case.

In this brief discussion, my intention is neither to take a stand for or against euthanasia nor to rest my case on one small article, but rather to engage with the rhetorical presentation of certain objections to Dr. Angell's testifying as these are detailed in the report: to do so both as a way into thinking about the modalities of care, concern, and advocacy I refer to in the title of this chapter, and as a way into considering some effects of an instituted social imaginary that holds them in place. Clearly, in so doing, I am accepting on faith the judgments of those who deemed Dr. Angell well suited, in herself and as a practitioner, to perform the required personal and professional functions that justify her agreeing to do just that. At issue are tacit, socially entrenched yet perhaps conflicting conceptions of specifically situated epistemic responsibility and agency as these characterize her practice. They tacitly shape the debates generated by this inquiry in particular, and they inform a wide array of often-contentious views about the place of advocacy and trust in the construction and public circulation of knowledge, now. In addressing these issues with reference to an "instituted social imaginary," I am again indebted to the thinking and practice of Cornelius Castoriadis, drawing on my elaboration of his work in *Ecological Thinking* and elsewhere in this present book. Recall that to the instituted imaginary Castoriadis opposes the *instituting* imaginary, by which he understands the critical-creative activity of a society whose autonomy is evidenced in its capacity to put itself in question; to recognize that, as a society, it is incongruous with itself, with scant reason for self-satisfaction.[3] This conceptual framing informs my thinking here.

Questions about *epistemic responsibility* had rarely figured in philosophical-epistemological deliberations until, with the development of social epistemology and the epistemologies of ignorance in the late twentieth and early twenty-first centuries, a conceptual space opened for

---

3. Lorraine Code, *Ecological Thinking: The Politics of Epistemic Location* (New York: Oxford University Press, 2006), p. 31. References are to Cornelius Castoriadis, *The Imaginary Institution of Society*, trans. Kathleen Blaney (Cambridge: MIT Press, 1998); and Castoriadis, "Radical Imagination and the Socially Instituting Imaginary," in Gillian Robinson and John Rundell, eds., *Rethinking Imagination: Culture and Creativity* (London: Routledge, 1994), pp. 136–54.

considering the issues they engage.[4] Yet ongoing resistance to according such issues philosophical legitimacy carries traces of a residual positivist-derived reluctance to allow that there could be multiple—and multiply valid—*ways of knowing* specific subject matters, events, objects, theories. Descriptively and normatively, the assumption has prevailed that knowledge worthy of the name will be univocal and definitive. Hence, as I observe in chapter 2, from the impersonal pronouncement "Science has proved" and its analogs favored by science journalists and other "outsiders," it appears to follow that "properly" objective knowledge could be anyone's or everyone's: it will, in consequence, claim universal validity. Since discussions of epistemic responsibility explicitly or implicitly invoke the figure of "the knower(s)" in her, his, their situatedness and concomitant particularity and/or fallibility, the concern has been that knowledge claims would be diluted or otherwise compromised in evaluative processes that invoke such responsibility judgments. In consequence, claims of/by particular knowers could fail to merit the (honorific) label "knowledge" or to achieve the level of certainty the label presupposes.[5] Yet claiming space and explanatory power for judgments of epistemically responsible knowing, whether individual or collective/communal, calls for more capacious assessments of cognitive achievements and methodological approaches than paradigmatic postpositivistic practices of verifying one knowledge claim against one item, event, or utterance in the physical-social-material world could allow. Nonetheless, as I propose in *Epistemic Responsibility*, evaluations such as these may well be required for accurately assessing the complexity and plausibility of knowledge claims that eschew abstract formality to return

---

4. See my *Epistemic Responsibility* (Hanover, NH: University Press of New England, 1987). Initially, the book was something of a sleeper, but the conceptual apparatus it introduces is currently claiming a place in postpositivist social epistemology and the politics of knowledge. See my "Testimony, Advocacy, Ignorance: Thinking Ecologically about Social Knowledge," in Alan Millar, Adrian Haddock and Duncan Pritchard, eds., *Social Epistemology* (New York: Oxford University Press, 2010), pp. 29–50.

5. See Barbara Herrnstein Smith, "The Unquiet Judge: Activism without Objectivism in Law and Politics," in Allan Megill, ed., *Rethinking Objectivity* (Durham, NC: Duke University Press, 1994), pp. 289–311. Pertinent is Smith's observation: "[N]o judgment is or could be objective in the classic sense of justifiable on totally context-transcendent and subject independent grounds. . . . [O]bjectivist claims may operate quite negatively under certain conditions and for certain members of the community and are in the long run perilous for the community at large" (p. 294). Such is the objectivism that condemns Marcia Angell.

to the world (individual and/or collective) of experience, experiment, and expertise, where all of these issues may figure in diverse processes of understanding and evaluation. In such revisioned inquiries questions, for example, about how the identities and circumstances of putative knowers may claim legitimate analytic/interpretive pertinence, as may the "why" and the "where" of any specific inquiry. There are choices about how, responsibly, to establish and to implement knowledge in situations, both scientific and quotidian, that are more multi-faceted, more variably textured than traditionally paradigmatic empiricist examples of knowing the cup is on the table, or the cat is on the mat, could have been.

In the inquiry at the Supreme Court of British Columbia where the disagreement that forms the substance of the 2011 article occurred, the lawyer for the Canadian federal government reportedly maintained that Dr. Angell should not be recognized as an expert witness "because she is an advocate for euthanasia and because her experience and training doesn't involve original research." Alleging that she is "passionate about [advocating for] assisted suicide"—in view of her having written articles in support of the practice—the government lawyer maintained that Dr. Angell had "sacrificed her impartiality" and was thus not capable of providing the objective testimony that the crown attorney could regard as valid, reliable. The implication is that *because* she cared, as an advocate by definition presumably would, she simply *could not* be sufficiently objective—be appropriately impartial, reliable—to present knowledgeable, well-informed testimony: her capacity to fulfill the obligations and expectations of an expert witness is thus subjected to radical contestation. In this scenario, advocacy as such, regardless of whose it is or how well informed the advocate(s) may be, finds ready, unreflective, condemnation infused as the criticism is by a presumption that caring deeply, unequivocally, damages a putative witness's capacity to know responsibly and "well enough"—even if that knower is an "eminent medical ethicist" closely affiliated with the relevant accredited institutions. There is also, I suggest, a further vital question about whether caring in such circumstances could actually be more problematic, more seriously damaging than caring "not a whit" or being coldly disinterested would be. The alternatives I present baldly, here, are integral to larger questions about advocacy and affect which, in the remarks cited, are cast negatively, as reprehensible, truth-inhibiting practices. Such a characterization is by no means neutral, morally, epistemologically, politically; nor is it convincing in itself, without further consideration.

Returning to my argument in *Ecological Thinking*,⁶ I continue to maintain, as I argue there, that advocacy can, in effect, make knowledge possible in the strongest sense of the word: in certain circumstances, it is a requirement *sine qua non* for the production, validation, and circulation of knowledge worthy of the label. Nor is it, without further ado, condemnable. These rather extravagant claims are not directed at "knowledge in general" which, I suggest, is an empty category. They are about knowledge pertinent to/in specific domains of inquiry and deliberation, where people who need to know well perhaps *cannot* (again, in a strong sense of that word) be expected to know for themselves by their independent, solitary efforts; and for diverse reasons. Would-be knowers may well be constrained, not for want of intelligence but owing to situation-specific barriers to their expertise, to their access and understanding: owing, perhaps, to the consequences of social-historical divisions of intellectual labor and/or entrenched structures of power and privilege as these are carried within the "received" epistemic imaginary, and hence circumscribe putative knowers' capacities to think toward new possibilities, away from fixed, sedimented ideas and expectations. In short, often, would-be knowers can neither find nor generate the requisite breaks in a putatively seamless epistemic imaginary that would allow untried ideas to claim a hearing; cannot practice or produce the "critical openness" that would allow them and/or their interlocutors to recognize their "biases and limitations" for what they are.⁷

Yet, although this thought will not be new to feminists, to science and technology scholars (STS), and/or to numerous other contributors to "postpositivist epistemology" projects, I am suggesting that the dogma of objectivity as it was invoked by the crown lawyer, in its starkest positivistic all-or-nothing construal, has not served "us" well.⁸ Some of the implications of this—perhaps outrageous—view bear rehearsing in connection with Marcia Angell's testimony, together with more wide-ranging issues

---

6. Code, *Ecological Thinking*, chapter 5.

7. José Medina, "The Relevance of Credibility Excess in a Proportional View of Epistemic Dependence: Differential Epistemic Authority and the Social Imaginary," *Social Epistemology: A Journal of Knowledge, Culture and Policy* 25, no. 1 (2011), pp. 15–35.

8. See also Lorraine Daston, "Baconian Facts, Academic Civility, and the Prehistory of Objectivity," in Allan Megill, ed., *Rethinking Objectivity* (Durham, NC: Duke University Press, 1994), pp. 37–63.

about the place of care, concern, and advocacy in multiple situations where it *matters* to know and to act responsibly, and well. Such places and situations are more common, more complex, and frequently more ambiguous than standard empiricist questions about how to know for certain whether the cup is on the table, or whether the barn facades that draw tourists to New England are fake or real. These more elaborated questions are germane to addressing the social-political implications and enactments of knowledge that involves, produces, or thwarts responsible social-epistemic recognition and interaction. My intention, then, is by no means to contest the value of objectivity in knowledge, from simple empirical claims to such convoluted questions as are at issue in the euthanasia case and in other analogous circumstances, but to urge more nuanced understandings of its scope and limits than have commonly prevailed, in Anglo-American philosophy, as in the wider world.

The idea that Angell has "sacrificed her impartiality" because she cares, because she is deeply "passionate about" assisted suicide, is at once so simplistic and yet so rhetorically definitive a dismissal as to close off space for (collectively) thinking more deeply about subjectivity, power, knowledge, and the place of care, responsibility, and concern in evaluating knowledge claims, expert opinions, and the ethics and politics of knowing. Whereas I present it here as a small moment in what will undoubtedly have been a fuller argument, such a flat refusal to accord epistemic respect to advocacy, for the reasons adduced, attests to larger issues in the politics of knowledge that require critical-constructive attention. On my reading of the report, Angell is being subjected to an egregious enactment of epistemic injustice in this exchange, in a sense distantly akin to the sense invoked by Miranda Fricker in her eponymous 2007 book.[9] Angell is impugned both in her capacity as a knower and as a trustworthy expert testifier: impugned on flimsy if not spurious grounds. Yet this condemnation finds support in a larger social-epistemic imaginary, precisely because of an entrenched—if sometimes warranted—*distrust* of advocacy, affect, and care, in twentieth- and twenty-first-century epistemology, and especially in scientific inquiry. Such distrust also finds support in the wider world where, as Lorraine Daston convincingly maintains,

---

9. Miranda Fricker, *Epistemic Injustice: Power and the Ethics of Knowing* (Oxford: Oxford University Press, 2007). Fricker characterizes testimonial injustice as "a kind of injustice in which someone is *wronged specifically in her capacity as a knower*" (p. 20), italics original.

still-current conceptions of objectivity require "not only freedom from theoretical bias but also a complete elimination of the personal and of the emotional."[10] Perhaps such strictures are loosening—for better or for worse—in the twenty-first century: the jury is still out as to their legitimacy and/or plausibility. But still in 2019, a conviction persists, if less starkly than was once assumed, that advocacy, with its undeniable specificity and consequent emotive components, can make no legitimate contribution to establishing the veracity of knowledge claims, nor to informing claims embedded in or intended to supply background for giving and receiving testimony. Despite its frequent aptness, distrust of this nature can neither be universally justified nor indiscriminately directed toward advocacy *as such*.

As I observe in *Ecological Thinking*, there are good public and private reasons to distrust crude, blatantly self-interested forms of advocacy *simpliciter* and, as I suggest in chapter 2 of this book, no need to rehearse the substance of its persistently negative reception. Nonetheless, some version of what I have called "taking subjectivity into account"[11] is surely required, case by case, in evaluating circumstances where expectations of trust and distrust prevail. The project would need to involve investigating at a deep (genealogical) level, not at a shallow superficial level, "whose advocacy is at issue, here?" (where "whose" may be singular or plural), and where a viable answer will need to resist *ad hominem/ad feminam* charges even in addressing the detail of specifically embodied "situatedness." Among pertinent specificities might be the credentials and epistemological "records" of would-be advocates, the social-political-historical positioning, genealogy, and conduct which confer presumptions of trustworthiness or its opposite upon their putative knowing. For Donna Haraway, whose conception of "situated knowledges" is a late-twentieth-century conceptual landmark,[12] such practices could—and in contentious situations often *should*—be enlisted to evaluate her, his, their qualifications for performing *these* acts of advocacy, here, in *these* circumstances.

Enlisting the conceptual resources of situated knowledges involves ruling out any prospect of achieving a view from nowhere (which, for

---

10. Daston, "Baconian Facts," p. 58.

11. "Taking Subjectivity into Account," in Lorraine Code, *Rhetorical Spaces: Essays on (Gendered) Locations* (New York: Routledge, 1995), pp. 23–57.

12. Donna Haraway, "Situated Knowledges," in her *Simians, Cyborgs, and Women: The Reinvention of Nature* (New York: Routledge, 1991), pp. 183–202.

Haraway, would count as a "god trick"), in order to recognize that knowing is always somewhere, and constrained and/or enabled by its situatedness. In consequence of such a conceptual shift, as I observe in chapter 1, inquiries may come down to dealing with particulars, but in responsibly deliberative processes they could do so without embarking on a pernicious slide into particularism.[13] The risk of such a descent, and of being dismissed accordingly, haunts many such projects. Definitive answers or resolutions may not be available in every case, but in eschewing the individualism and instantaneity of one-off, infinitely replicable propositional claims from which empirical knowledge before social epistemology allegedly was made, democratic deliberative inquiry can make room for the care-full investigations that engaging with matters of "individual" epistemic specificity, ethically and epistemologically, requires. These complex concerns about epistemic particularity/individualism cannot responsibly or adequately be judged, and/or dismissed, in simplistic "S has sacrificed the objectivity required to know p" discrediting. At issue, once again, is the very plausibility of "a view from nowhere": of a neutral, disengaged epistemic stance, especially in situations more clearly specific and situated than traditional cups-on-tables examples of putative epistemic neutrality have tended to be.

Nor is the requirement quite as simple as these comments may seem to suggest, for the report cited confirms that Dr. Angell's credentials *are* addressed, and *impugned*: her ethical-epistemological history, which in many quarters would be judged impeccable, is cited to discredit her. The discrediting may not be definitive, but it shows that even Angell's intellectual-professional eminence in the field of inquiry at issue is no guarantee against the egregious distortions on which advocacy's detractors, here, base their condemnations. The larger, if somewhat different, point is that deliberations about a respected practitioner's credentials must, almost by definition, be engaged *in media res*. They cannot plausibly be conducted from or on a perfectly clean slate, a *tabula rasa* cleansed of all traces of the epistemic imaginary that may have shaped the situations of their making, nor from a place where no preconceptions will infuse and color the debates. The disputes may reach no resolution; they may terminate in impasse. This is the stuff of which situated, real-world epistemic disputes are made: rarely do they lend themselves to the

---

13. See my "Particularity, Epistemic Responsibility, and the Ecological Imaginary," *Philosophy of Education Archive* (2010), pp. 23–34.

sanitized, hard-edged analyses formal Anglo-American philosophy of science and epistemology have offered, and on which orthodox positivism relies. Yet, the outcome need neither be a disdain for objectivity and for impartial principles of inquiry nor a slide into epistemic, ethical, or legal-political chaos. A democratic epistemic community worthy of the name will foster fair-minded deliberations about the nature and locations of expertise, and about the force of invoking matters of concern in an excessively authoritarian manner. There are few if any rules for designating fair-mindedness, but its absence can be palpable. Discursive practices for exhibiting its effects involve more care than the lawyer in this case appears willing to exercise.

Thoughts such as these, attributing certain epistemic shortcomings to "outsider" conceptions of the scope and limits of professional eminence, are implicit in the charge that Dr. Angell's concentration "on issues concerning pharmaceuticals" leaves her uninformed about matters related to euthanasia. Again, the claim is not trivial. It attests, in this aspect, to irresponsible advocacy on the part of her detractors, evident in their apparent failure to investigate or attempt to understand the complexities of this situation well enough to discern substantive overlaps within medical ethics—Angell's recognized and respected area of professional expertise—and to perceive the pertinence of the wide ranging questions that will have preoccupied her during her editorship of the *Journal* and beyond. The issue centers around public debates and judgments addressing issues about how "expertise" is to be reasonably conferred and recognized so that public deliberations about who rightly claims expertise, and about matters of care and concern more generally, could be fair minded in ways that the lawyer in this situation is failing to achieve. Tacit epistemological assumptions about knowledge as somehow filed away in hermetically sealed containers, where no overlap will occur from one "subsection" to another, attest to an implausible conception of human knowledge in general, and of medical knowledge in particular.

Even though there may be no good reasons to expect the government lawyer to differentiate within areas of specialization that are not her own, when it comes to disqualifying the testimony of an eminent practitioner and scholar in an open hearing, the public whose options will be influenced by the outcome expects more from credentialed practitioners by way of establishing why, in this situation, trust should be conferred or withheld. Such expectations—themselves vulnerable to misinterpretation or distortion—are integral to ensuring that the proceedings are informed by commitments to just, responsible epistemic practice. The claim is emphatically not that once it is determined that trust is appropriate in such an investigation, then "saying makes it so," even if the "sayer" is a recognized expert: this is no simplistic argument from authority, nor is the claim that testimony in a case of such eminence deserves greater care than in a case of lesser public prominence. Nevertheless, epistemically responsible practice is more likely to achieve emblematic status in so public an exposure, thus to prompt broader policy deliberations than routine, quotidian events might do. Crude categories and mechanisms of condemnation should be ruled *hors de question* in the absence of efforts to determine how well they "fit" situational specificities. On a different level, one consequence of eschewing an epistemic individualism for which isolated, discrete "statements of fact" are epistemologically paradigmatic, can be that patterns of verification and falsification will extend more widely and deeply than individually focussed, top-down inquiry presupposes. Single, isolated "one-liners" do not serve these purposes well. In a functioning epistemic-scientific community, difficult moral and policy questions demand discussion and negotiation, not instantaneous, univocal endorsement or dismissal, even in a courtroom situation where, in the interests of expediency, a (vacuous?) protest could be ventured to the effect that time is money.

To commend advocacy as a sometimes-legitimate—even laudable—epistemic practice and to claim a place for care and concern in knowledge making based on one small newspaper article would be flimsy indeed, inviting and warranting charges of epistemic frivolity. But Angell's contested positioning in the euthanasia case is exemplary beyond the courtroom setting, in ways that are also relevant for understanding the perceived situation of climate (and other) scientists in current twenty-first-century epistemic and political environments, where "manufactured uncertainty" prevails. Her situation poses urgent questions, even to participants who claim epistemic neutrality: questions about the ethics and politics of knowledge, and about professional expertise. They indicate some directions philosophical engagement with testimony and "expert" knowing has to take in moving away from the abstractions of positivism's formal modalities to engage with and attempt to understand real-world power-infused knowledge-making and -circulating practices, where science, epistemology, ethics, and politics cannot readily be disentangled.[14]

As I propose in chapter 2, addressing such questions "in situation" requires engaging with epistemic subjectivities in a manner quite foreign

---

14. Such is the thrust of Daston's argument in her "Objectivity and the Escape from Perspective," *Social Studies of Science*, 22, no. 4 (Nov. 1992), pp. 597–618.

to Anglo-American philosophy's persistent image of "the knower" as a disengaged, "remote" (following Val Plumwood[15]), interchangeable placeholder in the pursuit of knowledge, itself typically conceived as consisting of discrete, ubiquitously salient facts. Whether advocacy practices *should* participate in making knowledge possible or in contesting its claims—whether they should be accorded or denied epistemic respectability—will depend on who the advocates are, what credentials and justifications they supply for advocating as they do, how their trustworthiness is established or gainsaid in wide-ranging deliberative processes. Moreover, for purposes of this discussion and beyond, it makes good sense to take seriously the likelihood that reliable advocates (or informed dissenters), singly or as members of an advocacy group, may *care* about the claims and positions for or against which they advocate. This aspect of advocacy, in some of its modalities, commonly elicits condemnation; yet it is clear that in advocacy *tout court, caring* simply as such, in the abstract, can be neither applauded *nor* deplored. An equally troubling, by-no-means-far-fetched possibility is that, purely for personal benefit, people may advocate for views they neither care about nor believe to be true.[16] Such hesitations—such worries about potential conflicts of interest and hidden agendas—take on an acute urgency in times of political upheaval and the (often drastic) actions they incite. Nonetheless, how much a knower cares seems to have little bearing on her/his reliability. Returning to Angell, even from a commonsensical point of view it is difficult to imagine undertaking the frequently contestable and contentious work advocacy projects such as hers require, without caring about the outcome. *Ex hypothesi*, so to speak, advocates cannot routinely claim the impartiality—the principled "escape from perspective"[17]—allegedly integral to achieving objectivity "properly so called," ahistorically conceived and formally enacted, as it has infused the epistemic imaginary of the Anglo-American scientific, cultural, and philosophical mainstream since the Enlightenment. Again, I am suggesting that it is partly because she *cares* that Angell faces the criticisms that seek to disqualify her, as though it were, by definition,

---

15. Remoteness, as normatively characteristic of mainstream epistemology, ethics, and politics, is a theme in Val Plumwood, *Environmental Culture: The Ecological Crisis of Reason* (Abingdon: Routledge, 2002), esp. pp. 71–82.

16. Thanks to Jamie Robertson for this reminder.

17. The reference is to Daston, "Objectivity and the Escape from Perspective."

impossible to care reasonably, rationally, and knowledgeably. Because she *cares* as she does, her advocates have rallied to support her views and actions, whether wisely or otherwise: caring is not always truth-generating.

Adherents to postpositivist epistemological principles, likewise, view care with skepticism. Despite its promise as integral to responsible epistemic practice, for many feminists and other Others (from an invisible white masculine norm) in the Anglo-American world, *care* is a persistently double-edged, ambivalent concept and practice. Its warm, feel-good aura which has tended to situate it alongside "the feminine" as a naturally nurturing modality, contrasts with a darker side where women are confined as caregivers so as to enable the "serious" business of life to proceed, unencumbered by the onerous minutiae of domesticity involved in reproducing the work force; or where women are "cared for" in oppressive, paternalistic social-political structures and arrangements allegedly designed to protect them from the harsher realities of the world. The importance of care for knowledge production and circulation speaks to the implausibility of a commitment, among positivist-empiricists and practitioners of the natural sciences, to the view that a *bona fide* knower must approach her or his subject matter/object of inquiry dispassionately, openly, following where the evidence leads, regardless of the desirability or otherwise of the directions it takes or the conclusions it proffers. She/he should not *care* indiscriminately even if the outcome is unpalatable: she must keep her feelings in check. Flaws in this view are vividly apparent in Kristin Shrader-Frechette's showing how even "properly" objective inquiry can fail to take into account some of the most urgent issues peculiar to certain embodied, situated *subjectivities* which call for quite specific interventions, in view of the epistemic and ethical injustices consequent upon the routine invisibility of *these* vulnerable bodies in conclusions readily available to a nonscientific public, and to many scientists. It can amount to a call for care.

My purpose is neither to critique nor to endorse Shrader-Frechette's position. Rather, I propose working from it as a platform for engaging questions about the place of care in knowledge and the place of knowledge in care: questions about ignorance and/or about responsible epistemic-scientific practices, as these figure in feminist and other critical epistemological and moral-political theories. The purpose is to understand how they contribute to showing why sedimented assumptions about the autonomy of knowledge, about objectivity, epistemic agency, and the politics of knowing may have to be rethought when they start from the specificities of situated, vulnerable, perhaps extraordinary lives rather than from misbegotten convictions about human and locational sameness, interchangeability, and universal pertinence. These thoughts are not new in themselves, but they take on a renewed urgency when the language of care enters evaluations of knowledge production and circulation to perform a constitutive epistemic function, even if the caring involved cannot fit easily into the one-on-one interpersonal framework on which, for example, Carole Gilligan's and Nel Noddings's work on care in the 1980s tended to rely.[18] Thinking further about these issues, I again look to Shrader-Frechette, and especially to her thinking about the role of care in knowledge and of knowledge in care.

Many contentious testimonial and/or other knowledge claims merit elaborated, engaged analyses that depart from top-down "individual" spectator-epistemology practices of validation or falsification. Thinking and deliberating communally, ecologically, horizontally, across multifaceted situations, temporal locations, and circumstances may well be the best, perhaps the only way to establish or discredit the plausibility of novel or disruptive knowledge claims that unsettle the status quo and rely upon or contest the putative reliability of their would-be advocates. This is how it can be in situations where reputations of expertise are established or challenged: situations that call for more complex justificatory practices than do one-off, punctiform claims and counterclaims.

Requirements such as these are writ large in advocacy situations that engage—as Dr. Angell's does—with issues analogous to those Bruno Latour singles out as "matters of concern." Here I am reading his distinction between matters of fact and matters of concern more expansively than its initial presentation suggests, to make space for affirming close connections between concern and care, and affect. Latour deplores a tacit yet entrenched epistemological obligation to erase "the work required in order to establish the persistent, stubborn data . . . to limit 'facts' to the final stage in a long process of elaboration."[19] Such a requirement expunges genealogical traces from scientific and everyday knowledge-

---

18. Carol Gilligan, *In a Different Voice: Psychological Theory and Women's Development* (Cambridge, MA: Harvard University Press, 1983); Nell Noddings, *Caring: A Feminine Approach to Ethics and Moral Education* (Berkeley: University of California Press, 1984).

19. Bruno Latour, *The Politics of Nature: How to Bring the Sciences into Democracy*, trans. Catherine Porter (Cambridge: Harvard University Press, 2004), p. 95ff. (cited in my *Ecological Thinking*, p. 101).

seeking/making practices, confining them to the allegedly superseded "context of discovery," which disappears in the putatively real work that happens in the "context of justification." It discounts the interactive/*intra-active*, conflictual yet productive labor from which "factuality" is often achieved. Here, *intra-active* is Karen Barad's term: she employs it to work past a sustained separation between subject and object she finds typical of standard Anglo-American epistemic discourse: her purpose is to draw attention to the to-and-fro of epistemic talk where erstwhile "knower" and "known" remain distinct, separated throughout processes of inquiry: to close the gap. For Latour, by contrast, and germane to this discussion, "The only way to respect . . . heterogeneity and . . . locality is . . . to do a lot of philosophy. But philosophy is not about unifying factors . . . [It] is a *protection* against the hegemony of the present sciences."[20] As I show in *Ecological Thinking*,[21] questions about knowledge, responsibility, and agency are inextricably intertwined with the unevenly distributed cognitive resources and moral-political-affective effects of institutional knowledge production in present-day white Western societies. Practices of advocacy participate, albeit diversely, in shaping these interconnections, bringing such thoughts into public discourse, evaluating their implications for human lives.

Very Vulnerable Bodies

Thinking further about advocacy with respect to its bearing on a related aspect of the content, inspiration, and public/local impact of scientific-medical research, I will consider some implications of advocacy matters as they figure, if often implicitly, in Shrader-Frechette's 2007 book, *Taking Action, Saving Lives: Our Duties to Protect Environmental and Public Health*,[22] and in her ongoing research and practice. In the book, among

---

20. Bruno Latour, "Irréduction," in Werner Callebaut, *Taking the Naturalistic Turn; or, How Real Philosophy of Science Is Done* (Chicago: University of Chicago Press, 1993), p. 218.

21. Code, *Ecological Thinking*, pp. 60–61.

22. Kristin Shrader-Frechette, *Taking Action, Saving Lives: Our Duties to Protect Environmental and Public Health* (New York: Oxford University Press, 2007). Germane also is Kristin Shrader-Frechette, *Environmental Justice: Creating Equality, Reclaiming Democracy* (New York: Oxford University Press, 2002).

diverse matters of public concern, Shrader-Frechette documents the exceptional bodily vulnerability of children—conceived generically—to the effects of environmental pollution. She assembles scientific evidence with care and with a purpose: as an exemplary engagement with the kinds of advocacy practice this analysis defends, hers is an overarching, transparently caring and committed project. Questions about knowing well and advocating responsibly are pivotal in her analyses. Plainly, Shrader-Frechette is no neutral observer: she is deeply invested in—like Angell, she cares passionately about—the harms she addresses. Yet while she is no impartial bystander, her inquiry is objective in a strong sense with affinities to Haraway's "situated knowledges," which I cite in the Introduction, and to Sandra Harding's "strong objectivity."[23] For Harding, *strong* objectivity achieves its strength from taking the epistemic positioning of the scientist(s)/knower(s) into account in evaluating her/his/their knowledge claims. Its intent is not to cleanse inquiry of materialities or commitments, of interests, or of the constitutive effects of situation and place, but to analyze these as carefully as it analyzes typical "objects of knowledge." Its strength is in its self-reflexivity, its ongoing monitoring of its own processes of inquiry. Analogously, Shrader-Frechette's work achieves a significant level of objectivity *because of*, not in spite of her passionate commitment. Hers is no distanced, affect-cleansed, view-from-nowhere inquiry. As one reviewer aptly observes: "Shrader-Frechette's analysis is informed not only by her wide-ranging knowledge of relevant scientific material but also by her close familiarity with ethical theory. It is enlivened by a sense of indignation, compassion, and urgency."[24] Such is the affective positioning that makes certain kinds of empirical knowledge possible, while without compromising objectivity.

As Shrader-Frechette shows, knowing responsibly and well is required for fostering social justice in response to ecologically outrageous situations. The contrast is between knowing situations, persons, or things "in general," and knowing them well enough in their specificity and the variability of their detail to engage well with them—to "meet" them (in Barad's sense)—half-way. These are matters of epistemic and

---

23. Sandra Harding, *The Science Question in Feminism* (Ithaca, NY: Cornell University Press, 1986), esp. ch. 5. Here I draw on my reading of Harding in *Ecological Thinking*, pp. 61–62.

24. Hugh Lacey, review of Kristin Shrader-Frechette, *Taking Action, Saving Lives: Our Duties to Protect Environmental and Public Health*, in *Ethics* 118, no. 4 (July 2008), p. 761.

ethical responsibility. Because so few nonscientists are capable of knowing well enough without sensitively interpreted empirical data and engaged evidence of the kind she offers, they/we are reliant on advocacy such as Shrader-Frechette proposes if we/they are to doubt intelligently, to protest plausibly against recurring, authoritatively uttered public assurances to the effect that, environmentally-ecologically, things are "getting better."[25]

Countering "the longevity objection," whose adherents argue that "people seem to be living longer and getting healthier" in order to disarm charges that pollution damage to people's health in the USA is ubiquitous and increasingly dire, Shrader-Frechette remarks that the objectors "forget that people would be even healthier if pollution were reduced"; that "[r]egardless of longevity, environmental pollution . . . gives the most vulnerable people—children, particularly minority children—poorer health than they otherwise would have had."[26] Nonetheless, in state-of-the-art reports of pollution levels, the specificities of children's diverse vulnerabilities to a wide range of noxious substances are frequently invisible: they vanish into the numbers to count as just one person, one statistical unit among others. People tend not to think of taking them separately into account when they are routinely "included" as one unit in a set of statistics. Yet children's small bodily size, relative fragility, and general incapacity to look after or speak for themselves perpetuate widespread ignorance of their disproportionate susceptibilities to multiple events and illnesses that vary widely across race, class, social marginalization, and privilege. These matters can neither, responsibly, be left unaddressed, nor can they be contained within "one-size-fits-all" analyses. Shrader-Frechette's recognition that "we" need to *care* about these matters, to look behind the statistics to expose the gaps and exclusions that keep them in place, is a noteworthy achievement of the investigation, whose epistemic and ethical imperatives are interconnected and reciprocally constitutive. Statistical population analyses are poorly attuned to the specificities that, for concerned/caring inquirers—among them feminist epistemologists and other Others who depart from a positivistic, implicitly white male norm—require special attention.[27] Thus,

---

25. See Kyle Powys Whyte and Robert P. Crease, "Trust, Expertise, and the Philosophy of Science," *Synthese* 177 no. 3 (2010), pp. 411–25.
26. Shrader-Frechette, *Taking Action, Saving Lives*, pp. 32, 33, 34.
27. Consider Karen Messing, Barbara McClintock, and Rachel Carson, whose epistemological practice I discuss in *Ecological Thinking*.

to the assurance "statistics have shown," the question needs to be posed more pressingly, investigating what these statistics do not show or have not shown, what they are selected to show, what picture of social beings they work from and generate, and why "we" should care. Knowing well enough to advocate well in these and analogous circumstances requires engaging with diverse interlocutors, reading statistical evidence "against the grain," learning to recognize when "there is more to be said." There are many ways to read statistics. Here, informed advocacy is required to counter the epistemic and ethical-political injustices consequent upon the knowledge/ignorance (invoking Shannon Sullivan's term[28]) through which "facts" purporting to show that there is no danger from environmental pollution are circulated to a nonscientific public.

This said, David Michaels's insistence on the *negative* status and effects of advocacy clearly requires a response. It is a tangled issue. Referring to the tobacco industry's concentrated efforts to "manufacture uncertainty" so as to destabilize the "growing consensus linking cigarette smoking with lung cancer and other adverse health effects," Michaels contends: "Opinions submitted to regulatory agencies by corporate scientists and, especially, the product defense industry must be taken as *advocacy*, primarily, not as science."[29] Plainly, I reject so stark an opposition between advocacy and science. Hence, this contention both confirms *and* contests my position here. Given the rhetorical force and generous financial backing of campaigns to encourage people to keep on smoking, of which Michaels is deeply, justly critical, and given how "bad science" can be co-opted to sustain uncertainty about potential harms, thus promoting the agenda of the tobacco industry, his condemnation of these "opinions" is well warranted. They are, he observes, devoted more explicitly to advocacy enacted in the service of private, corporate interests than to scientific truth. Nonetheless, when science-informed claims from the "other side" are brought to bear, likewise animated by strong interests (even when they run counter to the agenda of the

---

28. See Shannon Sullivan, "White Ignorance and Colonial Oppression: Or, Why Do I Know So Little about Puerto Rico," in Shannon Sullivan and Nancy Tuana, eds., *Race and Epistemologies of Ignorance* (Albany: SUNY Press, 2007), pp. 153–72.

29. David Michaels, "Manufactured Uncertainty: Contested Science and the Protection of the Public's Health and Environment," in Robert N. Proctor and Londa Schiebinger, eds., *Agnotology: The Making and Unmaking of Ignorance* (Stanford CA: Stanford University Press, 2008), pp. 91, 102; my italics.

uncertainty pedlars), they too must be acknowledged as advocacy and evaluated accordingly. Here is the conundrum: the contrast is less between science and advocacy *simpliciter* than between epistemically *ir*responsible science informing advocacy conducted to serve harmful agendas, and advocacy informed by epistemically responsible science in the service of beneficial, ecologically sound social-political practices.

The dichotomy—between harmful agendas and responsible inquiries—is rarely as stark as this presentation suggests. When debates about smoking or the threats of climate change are conducted in "open" circumstances, the issues may seem to be clear, vociferous detractors notwithstanding; but antivaccination debates, for example, seem to occupy a space where no such clarity is available.[30] It is unclear which side is the "good" side, which side merits condemnation. In the climate debate or the smoking debate, matters seem to be clearer: pitting people's lives and health against profits and financial concerns will not do. In the antivaccination movement, it is less clear which side is "harmful." Vaccine advocates care about the health of individual children and of the general population: many people on both sides are worried about the health of their "own" children, even if their views may be less well informed. Criteria are elusive for judging one side to be harmful, the other beneficial. Science, loosely aggregated, can be used or misused on either side with clear, responsible directives remaining elusive. Yet epistemic responsibility has, at the very least, to play into deliberations about putatively harmful as contrasted with putatively beneficial agendas. Criteria have to be negotiated, deliberated, interpreted: such processes will be long and complex, but so it is with temporally and socially-geographically located knowing that fits uneasily with preestablished "truths." Too swift a conclusion would be epistemically, morally, politically irresponsible: failure to investigate further, likewise.

Practitioners of responsible advocacy will undoubtedly care about how their findings contribute to human health and well-being, and to reducing, preventing, and/or repairing damage to the physical-natural world; those whose relationship to knowledge is opportunistic and/or self-serving may care differently. But if advocates on both sides *care*, then even caring is not enough: not the decisive ingredient. More germane is a commitment to responsible knowing, where a "rush to judgment"

---

30. Thanks to Jamie Robertson for her contribution to these thoughts.

rarely merits applause. Hence, adjudication processes will involve intricate ethical-epistemic-political deliberations, not separated from but interwoven with advocacy-versus-science debates, for purposes of blocking descent into adversarial contests between conflicting camps, with matters of social justice and scientific "truth" falling by the wayside. These urgent issues resist premature closure. Undoubtedly, "the facts" have to be established; but if so doing reduces to a power struggle between advocacy informed by epistemically responsible science and advocacy funded by private-interest science, when the distinguishing criteria may also be unstable, the issue will remain fraught. Premature closure is neither epistemically nor morally viable. Nor need inquirers assume before the fact that private-interest research, by definition, abdicates allegiance to "the truth." As with the challenge to Marcia Angell, if differently in the detail of the issues, advocacy on behalf of "scientific findings" can be discredited, cast as dangerous because, in the going epistemic imaginary it is identified *only* with condemnable modalities of private-interest science and motivated interpretations of science in general, which compromise when they do not negate, the achievement of measured objectivity.

While no "one-size-fits-all" evaluation of advocacy projects is plausible, the alternative is not to opt for whatever interpretation seems, now, to fit. Care needs to be taken, communal deliberations and disputations undertaken and *time* allocated, to ensure that the judgments respond well to the specificities of situations and populations, horizontally and vertically and across intersecting, even contradictory, values. How such specificities should be known, what responses they rightfully elicit, requires engaged deliberations in extended conversations: collaborative, multiply engaged deliberations which could issue in action, even if they fall short of achieving certainty. There is scant reason to suppose that all such events will be calm, benign. Inquirers often have to settle for a compromise, a balance between tolerating a degree of ambiguity—of open-endedness—and of needing to act without delay. Outlining this potential impasse recalls Daston's astute observation: "[T]he most important factor in the ready acceptance of Baconian facts, despite their strangeness and/or irreplicability, was trust, extended almost *carte blanche* to at least a small circle of respected colleagues and informants."[31] If debate reduces to urgent and possibly fraught decisions about where to

---

31. Daston, "Baconian Facts," p. 49.

confer trust, how to avoid acrimony, then clearly these are as urgently matters of ethics, responsibility, humility and care as about matters of bare (empirical) fact. The issues reach well beyond academic philosophy: they pertain to widespread considerations of public trust, and of (secular) interpersonal trust, where no guidelines may be available analogous to those that govern public deliberations in institutional settings, too frequently characterized as adversarial. In the absence of such guidelines, and within the specificities of epistemic communities, the challenge is to determine honorable directives that prevent a descent into epistemic chaos. An epistemology based in empirically derived one-liners too often lacks the resources to curb epistemic chaos; yet a functioning epistemic community requires deliberative guidelines. How, then, to proceed?

Thus, for example, Shrader-Frechette's text can be read as a *condemnation* of advocacy in the strongest sense, as she records case after case of corporate advocacy bent upon "orchestrating ignorance, ignoring consent."[32] She shows, unequivocally, why advocacy has met with the condemnations I mention, as it was enlisted to discredit Marcia Angell and is enlisted against PR firms in the USA, "using selective facts and emotional appeals . . . to deny pollution problems . . . [or when] they cannot be denied . . . to show either that pollution produces compensating benefits, or that it cannot be stopped."[33] It is itself a work of advocacy, now positively understood: informed, eloquent, caring, engaged with matters of fact that inform and animate matters of concern, and vice versa: it makes a strong case for deliberative democracy and intelligent activism.[34] In the "right" circumstances and with informed, trustworthy participants, advocacy can make knowledge possible, knowledge that satisfies the highest epistemic standards and withstands the most rigorous scrutiny such as is essential for informed, justice-committed thinking and doing. Nor are these matters epistemological alone: clearly, they are at once ethical and political. Most urgently, they are also *ontological* matters that silently but insistently pose the question "Who do we think

---

32. "Orchestrating Ignorance, Ignoring Consent" is the title of chapter 2 of Shrader-Frechette's book.

33. Shrader-Frechette, *Taking Action, Saving Lives*, p. 64.

34. See Iris Marion Young, "Activist Challenges to Deliberative Democracy," *Political Theory* 29, no. 5 (Oct. 2001), pp. 670–90.

we are?[35] we who claim the power to make such decisions and to act upon them. Although easy answers may not be readily available, the question is not merely rhetorical: it figures prominently among urgent questions for putatively "first-world" thinkers and activists, now. Hence, Shrader-Frechette asks: "What ought citizens to do, to protect their rights [to know]? Instead of condemning all private interests—interests potentially essential to economic survival—citizens ought to use the tools of deliberative democracy to educate themselves and others, to participate in preventing conflicts of interest, and to ensure that government regulators and oversight agencies behave as they should." In her view, private-interest science is "neither scientific nor ethical," where the terms are to be understood both descriptively and normatively.[36]

Advocacy and/or Objectivity?

In social epistemology, where the disputes about Marcia Angell's testimony and Kristin Shrader-Frechette's analyses can helpfully be situated, and especially in *feminist and postcolonial* social epistemology, such knowledge claims as are at issue frequently draw upon narrated "real-life" examples to establish or illuminate their positions. A contrast might be with punctiform "one-liners"—with such examples, beloved of early empiricists, as knowing the cup is on the table, knowing the opening times of the bank: empirical "simples" which are testable in experience and are often found to be definitively true or false. Despite their promise, practices of drawing on such examples bring with them a set of challenges that expose inquiry to charges of "particularism," and to minimizing its exemplary reach in so doing. Such charges bring with them cautionary reminders, especially in quests for epistemic certainty.

More complex, in the twenty-first century white Western epistemic imaginary, is a propensity of story-telling—narrative—to slide into the anecdotal: into the fictional, contrasted with the (hard) factual; the incidental, contrasted with the necessary, the universal. For orthodox theories of knowledge, and *a fortiori* for moral knowledge, such specifically

---

35. I engage with this question in "Ecological Subjectivities, Responsibilities, and Agency," in Anna Grear and Louis Kotze, eds., *Research Handbook on Human Rights and the Environment* (Cheltenham: Edward Elgar, 2015), pp. 46–60.
36. Shrader-Frechette, *Taking Action, Saving Lives*, pp. 74–75.

situated narratives commonly invite epistemic distrust and may, in consequence be judged deeply flawed, epistemologically. Referring to moral knowledge, Cheryl Misak writes of a "feeling among some moral theorists that inquiry into what is right and wrong must be a rational or pure inquiry in which subjective experience takes a backseat . . . [indeed, that] moral experience is so contestable that it makes no sense to think of ethical deliberation as being a rational enterprise or to think that there is truth and objectivity in ethics."[37] Social epistemology loosens these strictures, opening spaces for engaging critically with the epistemic aspects of *testimony and trust* and for drawing on narrated, more affectively engaged examples than postpositivist epistemologists tend to favor. The shift is gradual, and not smooth. Still, it is from the detail of the narrative accounts that introduce her scientific/health care examples that Shrader-Frechette's statistical data claim a measure of power and of lay accessibility they might otherwise not achieve. Stories can bring damaged, suffering people into contact, into interaction—intra-action—with practices of other knowers and activists. In so doing, they perform a tacit advocacy function, inviting readers to *care*, showing them/us why, in *this* specific situation, care matters, while ensuring that "this situation" cannot be dismissed before the fact as an isolated, one-off, aberrant event. Thus Shrader-Frechette's book opens with the story of Emily Pearson, a child in a neighborhood east of Chicago, who developed brain cancer at the age of three and died when she was seven.[38] A different story introduces each chapter, grounding the accounts affectively and corporeally: humanizing them, advocating and demonstrating a responsibility to know, and to *care*. In its first chapter, the book traces patterns of childhood cancer, moving back and forth through demographic analyses and bystander stories of the effects of Emily's dying, and of other deaths, in a particularly polluted environment near the Ferro Chemical Plant. In the telling, Shrader-Frechette maintains a fine balance between narrative detail and statistical analysis, always to advance her advocacy project, but never losing sight

---

37. "Experience, Narrative, and Ethical Deliberation," *Ethics* 118 (July 2008), p. 617. Misak observes: "Medicine and health policy now resolutely take themselves to be 'evidence based' . . . they have turned to a narrow conception of evidence epitomized by randomized control trials . . . The dominant feeling is that the bad old days of subjectivity are gone and the new day of objectivity, enabled by EBM (evidence-based medicine) has dawned" (p. 616).

38. Shrader-Frechette, *Taking Action, Saving Lives*, pp. 3ff.

of its materiality or its specific poignancy; nor can she be charged with committing the philosophical sin of "generalizing from one particular."

The effectiveness of Shrader-Frechette's analysis derives, largely, from her respectful approach to testimonial—relational—events in twenty-first-century ordinary lives. Engagement with them affirms her responsible, knowledgeable concern for the harms she recounts. Their relationality manifests in the interpersonal trust whose integrity is sustained in ongoing personal interactions: in advocacy relations. The trust seems to animate an expectation of certainty in the inquiry's outcome: of certainty achieved, not manufactured; while the power of the analysis derives in large part from Shrader-Frechette's focus on affective, care-focussed aspects of situations where giving and accepting care requires trust. In view of the nature of the inquiry, it is reasonable to expect a degree of quotidian certainty to animate the caring that ensues, whereas in mainstream Anglo-American epistemology, individual knowledge claims, animated by care, are commonly denigrated for a putative failure of objectivity, a failure to expunge affect. Analogously, testimonial evidence is frequently a source of wariness, owing to its reliance on specificities. Recall the traditional ordering of perception, memory, and testimony as sources of empirical knowledge: testimony ranks third not just numerically, but for its third-place valuation: its "distance" from the facts. In postpositivist empiricist theories of knowledge, testimony-derived claims are contestable owing to a wariness of their inevitable specificity and situatedness, and the risk of compromising their objectivity. The contrast is puzzling.

Two examples that engage analogous issues, if from quite different intellectual places and conceptual situations, are worth noting here. Consider, first, the fictional character Alexandrovich, in Leo Tolstoy's *Anna Karenina*, who is commissioned to investigate the lives of "the racial minorities," and second, John Hardwig's sequentially and anonymously populated experiment where he investigates the implications of epistemic trust for a group of collaborating physicists. Both accounts engage with socially situated practices: contrasted with Shrader-Frechette's examples, both are formal in their detachment from the significance and effects of knowing, yet because both, albeit diversely, give content to the idea of "epistemology without a knowing subject"—a Popperian desideratum—and because in neither example does the "knowing subject" appear, it comes as no surprise that, in consequence, these framings make no space for affect—for caring—as it may have contributed to the epistemic activities where the claims originated, even if only to disdain them. Since omissions of affect are rarely noted in analyses of Anglo-American epistemic practices, these omissions are neither condemnable nor laudable. They conform to an epistemic orthodoxy for which neutral, affect-free objectivity is *sine qua non* for the very plausibility of knowledge. In this, as in other ways, they contrast starkly with the knowledge making central to Shrader-Frechette's project. Given the ordinariness of disengaged epistemic practices in analytic epistemology, it may seem bizarre to mention them in relation to affect: standard analytic theories have no place for it. Nonetheless, affect must have played a part in the advocacy integral to Shrader-Frechette's examples, in view of the specificities of their recounting. The differences gesture toward the epistemic cost of its absence. Advocacy's absence from the Tolstoy or from the Hardwig example, with the affect that commonly fuels it, might well be read as closing inquiry off from places where there are no prospects of manufacturing certainty *or* uncertainty.

Consider the following: In *Anna Karenina*, Leo Tolstoy writes:

> The new commission . . . investigating . . . the life of the racial minorities was appointed and sent to the scene with extraordinary swiftness and energy . . . The life of the minorities was investigated in its political, administrative, economic, ethnographic, material and religious aspects. *All questions were furnished with excellent answers, and answers not open to doubt, since they were not the product of human thought, which is always subject to error, but were the products of institutional activity.* All the questions . . . about why there were crop failures, why the populations clung to their beliefs . . . now received a clear and indubitable resolution.[39]

Here, institutionally produced knowledge functions as the arbiter of truth and facticity, whose trickle-down effects in people's lives play a constitutive-normative part in shaping the social order they inhabit. Contrary to my plea for the (implicit) trust and trustworthiness integral to the uptake and communication of empirical knowledge, Tolstoy's appeal to the indubitability of "impersonal" institutional knowledge—to its

---

39. Leo Tolstoy, *Anna Karenina*, trans. R. Pevear and L. Volokhonsky (London: Penguin Classics, 2004 [orig. 1873–7]), p. 370.

capacity to block issues of care and concern—tacitly applauds this erasure of testimony and trust. The words/thoughts of testifiers-informants are removed from scrutiny and accountability alike. Some might contend that Shrader-Frechette's examples are vitiated by the opposite conditions—by their too specific, too personal connections: this is the issue. The Tolstoy excerpt is fictional, if no more fanciful than questions about fake barn facades as openers into deliberating about who knows, and how. Even in its elaborated form, it retains a facelessness that points toward the complexity of large textual issues in clarifying the "place of knowledge in "our" social cognitive ecology," as contrasted with standard S knows that p assertions. Like them, the knowledge Tolstoy considers is affect-free, affect-neutral. Epistemic subjectivity and agency disappear, as knowledge becomes a distributive commodity, and questions about why a knower should care disappear from view.

For Alexandrovich, the commission's answers were not open to doubt because *"they were not the product of human thought, which is always subject to error"*: a trajectory that stretches from officials and priests to regional superiors and vicars to governors and bishops, with God and Mammon in accord, is the descriptive and normative source of epistemic *certainty*. In the indubitability conferred by the dispassionate public facade of their offices, Tolstoy's investigators do not count as identifiable knowing subjects (*vide* Popper). Questions about their credibility or about the genesis or warrant of their knowledge and credibility would be out of order, for the context of discovery is closed to scrutiny by fiat; there is no space to interrogate the processes of justification or to consider why such knowledge matters or who cares. The circumstances of the cognitive ecology—of its genealogy, in Michel Foucault's sense—confer upon it a curious presumption of infallibility. Yet there is another Popper who maintains (in *The Open Society and Its Enemies*) that "objectivity is closely bound up with the *social aspect of scientific method*"[40] and whom Helen Longino reads as vacillating between this starkly asocial ideal of "epistemology without a knowing subject," and promoting a social conception of criticism.[41] An asocial presumption, I suggest, also shapes the excision of affect from Tolstoy's account. Longino

---

40. Karl Popper, *The Open Society and Its Enemies* (London: Routledge, 1945).
41. Helen Longino, *The Fate of Knowledge* (Princeton: Princeton University Press, 2002), pp. 5–6.

discerns an epistemological challenge in the vacillation, "of reconciling the cognitive rationality of the sciences with the conditions of their practice": a challenge central to her project of assessing "the fate of knowledge" after the naturalistic turn. The Tolstoy passage opens space for engaging a challenge of this nature, in the "institutional activity" that has produced the "official data," and also, if circuitously, in asking *who cares*, and why, in this starkly impersonal tale. Read against the grain, the passage is about chains of public knowledge and authority constructed in and acquired from testimony: about the conditions of their provenance and warrant. I read it through a lens focused on trust and authority, and on the practices that produced this "clear and indubitable resolution," to evaluate their social-ecological dimensions. Why should/could they deliver indubitable knowledge? Whose interests, whose well-being, whose concern and affect are fostered in or occluded by these findings? Who would place trust in them? Who would care? Diverging in the metaphysical assumptions that inform their thinking yet converging in aspects of the epistemic imaginary that animates these imaginings of knowledge and certainty, Popper and Tolstoy expose some social implications of these conceptions of rationality in ways that, with variations, preoccupy and puzzle would-be knowers still now. Striking to a twenty-first-century reader is Tolstoy's laudatory assertion that "the answers were not the product of human thought." If this claim is plausible, then it is not clear how readers are to respond to errors or anomalies in the chains of inference that contributed to these putatively "excellent answers"; nor is it clear why a knower should care to know. The excerpt suggests that knowledge can establish itself—temporarily—in a social-political space where situationally specific matters are germane to its intelligibility. Enlisting it calls attention to shortcomings in Hardwig's analysis, where the imperviousness he claims for the report removes the context of discovery from scrutiny. It likewise closes spaces for addressing the affective import of asking: "Who cares?" The indubitability of the inquiry's "resolution" speaks of a social cognitive ecology more closed than Hardwig's community of scientists could be, while portraying the commission as unassailable in yielding valid conclusions, based in expertise and trust. So "closed" a stance is integral to some iterations of logical positivism: my purpose is to show its limitations. Ironically, it carries an inherent vulnerability in consequence of the fragile unassailability that attaches to an epistemic ecology of putative certainty: it needs to be approached with care.

Initially, Hardwig's essays were innovative in affirming a constitutive role for cooperation and trust in science, and in contributing to a shift toward understanding science and knowledge-seeking more generally as social practices, contrasted with the inexorable processes of formal, impersonal discovery and justification that had dominated Anglo-American epistemology, as is announced in Karl Popper's famous dictum: "Knowledge in the objective sense is *knowledge without a knower*: it is *knowledge without a knowing subject*"[42] with its brief for the definitive veracity of knowledge claims uncontaminated by human thought and closed to prospects of naturalizing or socializing inquiry. The traditional epistemological emphasis on the cognitive practices of "individual knowers" is a slightly less troubling epistemological picture. But, Hardwig's work is much closer to recognizing, as Melo-Martín and Intemann remind us, that significant areas of present-day science "require the participation of large national and international teams of researchers with multiple forms of expertise, which ranges from statistical approaches, to instrumentation, to specific scientific knowledge in particular sciences."[43] On Melo-Martín's and Intemann's analysis, the figure of the individual knower does not, I think, lose all relevance. But, when epistemologists take institutional/scientific knowledge as their principal exemplar, many of the assumptions that govern philosophical engagement with "the knower" will have to revise their fundamental terms of reference. This thought is not intended to devalue individual knowledge-making, but to recognize the parameters of its reach, and to depart from any assumption that knowledge is paradigmatically a product, or a multiple, of *individual* knowing(s). In consequence, the politics of knowledge now is complex, vibrant, and more challenging (in a positive sense, I suggest) than standard "S-knows-that-*p*" exemplars were.

The tenacity of the ideal of the individual, or even of the absent knower, as guiding principles in knowledge seeking begins to account for the difficulty of introducing affect and care into epistemic projects, then and still now. While Hardwig's analyses of science as a social enterprise involving trust attempt to overcome the limits of an individualistic

---

42. Karl Popper, "Epistemology without a Knowing Subject," in his *Objective Knowledge: An Evolutionary Approach*, rev. ed. (New York: Oxford University Press, 1979), p. 109.

43. Inmaculada de Melo-Martín and Kristen Intemann, *The Fight against Doubt: How to Bridge the Gap between Scientists and the Public* (Oxford: Oxford University Press, 2018), p. 55.

portrayal of knowledge-making, his claims exceed their explanatory limits when they are enlisted to address social knowledge in a cognitive ecology where testimony in multiple, diverse modalities has become a principal epistemic resource. Nor do they open into innovative ways of engaging with epistemic authority beyond the confines of professional scientific communities.

Crucial for considering the implications of Hardwig's example is how the participants whose names are attached to the experiment are linked by trust in their credentials as *bona fide* scientists equipped with the relevant qualifications. Their epistemic responsibilities are circumscribed by the experiment's formal dictates, and by the reach of their expertise. Because none of them *can* (in a strong sense) verify the results of each stage for themselves, they are epistemically interdependent all the way down: any rupture in the line of trust, any short cut or fudging of results, would jeopardize the entire sequence. Thus, despite their being specifically named in the list of participants, the question persists: "Who really knows? Whose knowledge is at issue here?" For methodological individualists, it is unsettling: the answer, it seems, must be "anyone, and everyone." In fact, the "interdependence" Hardwig attributes to these contributors is bizarre in its formality, in the sense that they may well be interdependent, but there seems to be no communication, no interaction, between or among them. They are placeholders in a chain of contributors, demonstrating the effectiveness of a linear epistemic event. How their participation contributes to understanding the role of trust in knowledge is bizarre in the attenuated conception of trust that appears to inform it.

In principle, the expertise and epistemic authority of the scientists/ experts listed by name could be called upon to "double-check," to revisit their findings, were a weak link in the explanatory chain to be suspected. Listing their names offers a tacit warrant, an unspoken "trust me" assurance that is concealed from view by the cloak of anonymity surrounding Tolstoy's commissioners.[44] Presumably, real human thinkers/ knowers had produced Tolstoy's indubitable answers; but claiming that

---

44. In like vein, Elizabeth Fricker asks "whether the supposed ideal figure of the autonomous knower, who refuses ever to trustingly accept another's testimony, *a fortiori* will never allow her own judgement to be corrected by another's, is really such an ideal after all." Fricker, "Testimony and Epistemic Autonomy," in Lackey and Sosa, eds., *The Epistemology of Testimony*, p. 239.

these "were *not* the product of human thought" attests to the power of a social-epistemic imaginary where investigating and reporting could occur, and epistemic authority could be established, only *in absentia*. Hence, it is unclear how or where trust could rationally be placed, unclear how the epistemic authority of the investigators could be confirmed or contested. So, although the participants in Hardwig's experiment are named and thence "personally" accountable, their scientific expertise in contributing to the experiment and in using the apparatus that, for purposes of this experiment, were *theirs*, is presupposed. Such is the locus of their answerability. In the inquiry as Tolstoy depicts it, were the participants named and held singly accountable, or accountable as participants in a hierarchal division of epistemic labor, a more diverse, politically freighted range of questions, capacities, inclusions, and exclusions could be exposed to scrutiny. Such interventions would initiate a shift toward the engaged openness that democratic social-political epistemic practices seek to enable. Here, the issue is not about physical science transcending politics but about how, in this discussion, it is less *visibly* political; the questions it generates are qualitatively different from those the Tolstoy inquiry poses.

In proposing this contrast with the Tolstoy example, I am claiming greater significance than some commentators might for *naming* Hardwig's contributors, hence for viewing them as individually accountable for pieces of the whole, and for the relations of trust and concern they as named participants must, if tacitly, honor. Still, I do not mean to ignore the urgency in Hardwig's asking: "Does anyone know that *p*? Is that *p* known?" The question, he says, is disturbing "because it reveals the extent to which even our rationality rests on trust, and because it threatens some of our most cherished values—individual autonomy and responsibility, equality and democracy."[45] Read through investigations in naturalized and social epistemology, this question points toward acknowledging the limitations of individualist thinking, and toward initiating a rearticulation of those "cherished" values and bringing epistemological projects down to earth. Rather than representing a threat to democratic values, this opening is commendable in liberating epistemological inquiry from the unrealizable demands of an overblown autonomy ideal, and

---

45. John Hardwig, "Epistemic Dependence," *Journal of Philosophy* 82 (1985), pp. 335–49; and "The Role of Trust in Knowledge," *Journal of Philosophy* 88 (1991) pp. 693–708.

from what Guy Axtell aptly calls a "platitudinous adherence to epistemic value monism"[46] consequent upon a too-narrow focus on propositional knowledge claims.

The Tolstoy example, I suggest, exposes the epistemic underpinnings of an institutional authority that differs from the authority Hardwig's example must assume, in its (i.e., the Tolstoy example's) presentation as a "case" closed to any possibility of asking whether "anyone knows that p?" Now, in the twenty-first century, re-reading Tolstoy's "project description" against the grain, for its ironic meaning, calls for a quasi-genealogical inquiry into how that specific context of discovery could have generated the results he announces. The impact of naturalist-social inquiry for epistemology has been to create spaces for posing those very questions, as feminists and other Others have shown who eschew the epistemic individualism which, Hardwig notes, "dies hard."[47] These spaces open into pointed questions about whose knowledge "we" (epistemologists) are talking about; about the place of testimony *and of care*, in the construction and circulation of knowledge, with its part in forging or reconstructing epistemic community. Comparing these examples so long after the fact calls for examining how epistemic power and privilege differentials shape, and skew, patterns of commonality, interpretation, affect, and trust. The apparent ease of appealing to mutual trust in the Hardwig example contrasts with an uncertainty about the wisdom of taking trust for granted in the Tolstoy example, where there is no space for the critical discursive interactions that, according to Longino, are defining features of social knowledge.

Here I am referring to testimony widely conceived, to cover a range of possibilities from exchanges of knowledge-as-information—A knows that p, tells B that p and B, thus, justifiedly claims also to know that p—through to socially enacted transmissions and acquisitions of knowledge from conversations, debates, books, teaching, media, newspapers, lectures, films and other communications; thus also to the published record of Hardwig's experiment and Tolstoy's report. This take on testimony falls roughly into a nonreductivist camp in minimally maintaining that "one" (singular or plural) can come to know directly from testimony *when*

---

46. Guy Axtell, "Expanding Epistemology: A Responsibilist Approach," *Philosophical Papers* 37, no. 1 (2008), pp. 51–87.
47. Hardwig, "Epistemic Dependence," p. 339.

certain conditions to do with the reliability and responsibility of testifier(s) and hearer(s) are in place. The caveat is significant: such conditions routinely hold only for relatively simple "naive empiricist" exchanges where one could imagine, aptly, that the speaker and hearer are communicating on/from a level playing field, and incommensurable material-social power and privilege distributions can be taken for granted, as they cannot in Tolstoy's narrative. Matters of care find no place in investigations such as his. But the extent to which they can be assumed to inform the Hardwig example also requires re-assessment. It is not a given that information-conveying exchanges "telling-B-that-*p*" are paradigmatic of "things," "truths" people know through testimony, whether in philosophy or in the everyday world. Complex testimonial exchanges cannot, routinely, be understood as multiples of or derivative from these simple knowledge-conveyings. Nor can would-be knowers assume it is easy to discern where trust can rationally be placed. Ironically, the appeal of an "epistemology without a knowing subject" is clear in these thoughts: it would avoid the messiness of personal situatedness, trust, and caring—indeed, of human lives!

I have been working with a benign conception of testimony and trust in Hardwig's experiment and in everyday testimony in ordinary epistemic exchanges. Relations of testimonial trust, the account assumes, connect Hardwig's participants where, for argument's sake, I leave aside the problem of vulnerability that looms large for testimony's detractors. It seems to inform Hardwig's worry about a loss of individual autonomy. He notes, "[T]he trustworthiness of members of epistemic communities is the ultimate foundation for much of our knowledge."[48] If the observation is intended as a remark on an assumption of plausibility, then it rests on a fragile social/ethical basis. As indeed it must. While questions about the fragility of trust may be pursued about aspects of scientific practice, with competitiveness and personal reputation figuring prominently, in the Hardwig example "secular" trust prevails, and is maintained by justified, relatively durable relations of trust. As for Longino, knowledge production, like the production of results in Hardwig's example, is presented as having accumulated "through a community's continued engagement in a set of practices guided by a given set of standards." There "the epistemic dependence of scientists on one another . . . is a common,

---

48. Hardwig, "The Role of Trust in Knowledge," p. 694.

not an anomalous feature of scientific inquiry): common standards stand as safeguards against the fragility of trust and the vulnerability of the trusting."[49] *Absolute* trust, not unlike *absolute* truth is indeed ephemeral.

For Tolstoy, trust is of the essence for so confident a claim as his, about the inquiry's results, to be advanced, even if the people reliant on them are variously vulnerable. Differences in power, expertise, and authority in the contributors' rankings are as significant as unmarked membership in a linked chain of scientists, and not necessarily to the trust-enhancing benefit of the findings. While something like an *epistemic principle of charity* might inform a reading of Tolstoy's report—an assumption that the contributors' epistemic authority is "more or less reliable"—it is not clear that such a presumption could hold in situations of social, cultural, political invisibility and/or blatant inequality. In a well-functioning democracy, it is plausible to assume an *internal* trustworthiness in such an enclosed scientific community. And yet the Hardwig example, with the exemplary status it accords to *epistemic* interdependence, has less to offer a democratic epistemology than the Tolstoy example does, despite the latter's showing by negative example how to pave the way for the proto-democratic social order Tolstoy's Levin hoped to animate.

Briefly, then, the experiment prompting Hardwig's analyses claims an explanatory power that exceeds its limits; nor is it plausible to read the putatively egalitarian epistemic-scientific community he draws upon as a model for social and epistemic interdependence as such, or for democratic epistemic views writ large. Sandra Harding's contention in *The Science Question in Feminism* is germane. She claims: "a critical and self-reflective social science should be the model for all science, and . . . if there are any special requirements for adequate explanation in physics, they are just that—special." She continues: "Whereas the social sciences must consider physical constraints on the phenomena they examine, the objects, events, and processes of concern to physical scientists are limited to those that can be isolated from social constraints."[50] Subsequent feminist and other postcolonial analyses of social constraints in the physical sciences have shown these limits to be less hard-edged than they appeared to be in 1986, yet Harding speaks indirectly to the truncated relevance of the Hardwig example for social epistemology, after all. Her view could support

---

49. Longino, *The Fate of Knowledge*, pp. 150, 153.
50. Harding, *The Science Question*, pp. 44–45.

a suggestion that the Tolstoy example *with* its instructive shortcomings offers a hermeneutically and ecologically more promising route into these matters than Hardwig's, largely because Tolstoy's scenario is populated, situated, *in media res*, where resources can be found for confirming the human effects of the knowledge-making.

Read side-by-side, Hardwig's experiment seems to be the more transparent of the two: more open, more immune to critical scrutiny, whereas the screen of anonymity behind which Tolstoy's "knowers" know invites wariness, caution. Still now, in the twenty-first century, Tolstoy's exclusion of the knowing subject(s) from knowledge production asks for explanation, at least for thinkers who move away from the confines of postpositivist epistemology. Hardwig's example presents a picture of seamless, benign trusting relations where credibility is readily assumed and conferred: in that too-easy assumption it invokes caution. Neither example addresses the epistemic *subjectivity* of the knowers, who presumably *care* (benignly or otherwise) in seeking/claiming to know. Hardwig has taken on faith the trustworthiness of the testifiers on whose findings the experiment relies. Hence, he need not consider such factors as affect, bias, and error, nor of willful deception, which cannot be ruled *hors de question* in evaluating the testimony even of quasi-anonymous participants in carefully monitored inquiry. Perhaps because of the demographic homogeneity of the community he relies on, Hardwig works from a default assumption of the testifiers' uniform trustworthiness (and willingness to trust). In this self-contained situation, this assumption may in fact be plausible, well warranted; but in voicing concerns about individualism and democracy, Hardwig neglects to consider how far, for example, according to Kristina Rolin, "trust in scientific testimony involves trust in the community's ability to facilitate inclusive and responsive dialogue based on shared standards of argumentation."[51] For social epistemologists now, such questions are central, germane to the esteem of epistemic contributions.

Of epistemic exclusion, Nancy Daukas aptly observes: "In a culture plagued by racism, sexism, and other forms of oppression, most community members have internalized, to varying degrees, pernicious stereotypes regarding the cognitive and moral traits of socially constructed

---

51. Kristina Rolin, "Gender and Trust in Science," *Hypatia: A Journal of Feminist Philosophy* 17, no. 4 (Oct. 2002), p. 106.

'kinds' of people."⁵² The patterns and perceptual-epistemological habits that perpetuate such putative "knowing" tend to be enacted in a social imaginary which carries the often-implicit systems of social meanings and interlocking explanations-expectations within which people enact their knowledge and subjectivities and craft their individual and/or collective self-understandings and social-political expectations. Such systems cannot be presumed bias-neutral; yet neither are all so-called biases pernicious. An imaginary, recall, is self-reinforcing, much as self-fulfilling prophecies are. Ongoing "successes" consolidate a sense of its rightness: it is tenacious in protecting that "sense" from critique. An imaginary is neither contained by the regulative assumptions of normal science, nor is it only about how knowable items in historical periods and disparate places are spread out before "the" observant knower. So when Daukas emphasizes the *epistemic* significance of socialization within the norms that guide epistemic practices—such as stereotypes of "human kinds"—she is showing how stereotypes embedded in a received imaginary "work their way into . . . background presuppositions" where certain "kinds" of people will be deemed competent, epistemically and socially (or not), across a range of situations, activities, and patterns of judgement."⁵³ Such patterns pertain without contest in the Tolstoy example: they inform its conclusions about why their crops failed, why they clung to their beliefs, and thence how, socially and politically, they should be treated and trusted in light of these findings. Yet they do so within a necessarily limited, even local, social epistemological structure.

In the public imagination that informs a settled, homogeneous social order, as I note with reference to Marcia Angell, advocacy is routinely charged with egregious epistemic violations: with a failure of objectivity where vested interest overrides appropriate detachment; with abusing power and violating the integrity of the inquiry itself; with initiating a pernicious slide into relativism. But grouping advocacy practices together as violating truth and responsibility standards damages possibilities of engaging in the informed, caring, knowledgeable advocacy integral, for example, to the inclusive democratic activism Iris Marion Young was articulating in her

---

52. Nancy Daukas, "Epistemic Trust and Social Location," *Episteme* 3, nos. 1–2 (June 2006), p. 114.
53. Daukas, "Epistemic Trust and Social Location," pp. 115–16.

final writings. She contends: "An appropriate conception of democratic communication should reject . . . [the] opposition between reason and emotion . . . emotional and figurative expression are important tools of reasonable persuasion and judgement."[54] The urgency of Vrinda Dalmiya's question, "Why should a knower care?"[55] confirms how deeply entrenched (in affluent Anglo-American societies) is the presumption of detachment, remoteness, consequent upon normative-prescriptive separations of reason from affect, which are awkwardly derived from a conviction that affect sullies reason but with no counter assumption that reason could, *intra-actively* (again borrowing from Karen Barad) inform affect and make it wise.

Elsewhere, I have proposed that knowing other people might serve as a model for knowledge/knowing. In its basic (if often unspoken) commitment to knowing people in their specificities and differences across multiple instantiations, this is a model and a practice more amenable to articulating and enacting the diverse requirements of responsible inquiry than is knowing the medium-sized material objects[56] that are the focus of Anglo-American epistemic exemplars (knowing that the cat is on the mat, for example). In revisiting these thoughts, I do not follow Dalmiya in seeing her question as applicable, primarily, to knowing other people in dyadic one-on-one relationships: I read it broadly within a cognitive-affective framework where caring is part of the process: it is widely and subtly conceived, to catch a sense of cognitive *engagement*. The project affirms a transformative potential for care, consistently thwarted by an imperative that drives scientific and secular knowledge toward law-like conclusions and grand generalizations, aggregating particulars and blocking ways of knowing them in the specificities of their effects and *meanings*, and in their intra-actions. Acknowledging that the matters of fact Shrader-Frechette documents are just as significantly matters of care and concern, and that these concerns go to the roots of who and what "we" are as social-political human beings, such examples confirm the

---

54. See Iris Marion Young, *Inclusion and Democracy* (Oxford: Oxford University Press, 2000), p. 39; and "Activist Challenges to Deliberative Democracy," pp. 670–90. Shrader-Frechette looks to Young's work on deliberative democracy for the conceptual apparatus that informs the ethical implications of her analysis.

55. Vrinda Dalmiya, "Why Should a Knower Care?" *Hypatia*, 17, no. 1 (Jan 2002), pp. 34–52.

56. See Lorraine Code, *What Can She Know? Feminist Theory and the Construction of Knowledge* (Ithaca, NY: Cornell University Press, 1991), especially chapter 2, "Second Persons."

reach of *care* as a vital component of inquiry, even as it (wisely) limits prospects of achieving certainty.

Conclusion: Knowledge as Collective/Collaborative Practice

For social epistemologists of the late twentieth/early twenty-first centuries in the affluent Western-Northern (predominately white) world, John Hardwig's work is influential for introducing a range of puzzles and conceptual questions centered around matters of certainty: matters largely absent from Anglo-American epistemology prior to the mid-1980s. They contrast, intriguingly, with the "uncertainty" that is a central theme of this book. Departing from the demographic homogeneity and epistemic univocality of Hardwig's 1985 analyses, I have proposed considering an instructively different example that informs a fictional, yet plausible example of authority in institutional knowledge, in Leo Tolstoy's novel *Anna Karenina*. Such examples point toward an imperative for social epistemologists to reconsider trust as it is implicated with authority, responsibility, and epistemic transparency. Admittedly, the epistemic trajectory in the production of Tolstoy's "indubitable answers" is only distantly analogous to Hardwig's analyses: not only because the Russian inquiry is distant in time and place from Hardwig's, nor because it is fictitious, but in consequence of differences between social scientific and physical scientific practices, both then and now.

For Helen Longino, Hardwig's "puzzlement" could be dissipated by rejecting suggestions that, whereas the group as a whole knows, no individual within it does: by rejecting claims to the effect "that it is possible to know vicariously." As a way around this impasse, Longino proposes a distinction "between knowledge production and knowing," suggesting that while the group may have *produced* the knowledge, with "critical interactions at various stages and . . . critical discussion of the whole experiment,"[57] once the experiment is completed and the report

---

57. Longino, *The Fate of Knowledge*, pp. 152–53. According to Longino, the earlier Popper "is often considered to have originated social epistemology with his emphasis on the importance of criticism in the development of scientific knowledge" (pp. 5–6). See also Rolin, "Gender and Trust in Science"; and Fricker, *Epistemic Injustice*, p. 112 n7. Hardwig's work is cited in Lackey and Sosa, eds., *The Epistemology of Testimony* as a pivotal text in social epistemology.

is published, she maintains, the knowledge is *commonable*. Yet, in effect, the suggestion offers little more than an appeal to commonability, with a reminder that people inside and outside an experiment or other inquiry can claim, rationally, to know what it makes publicly available so long as they do sufficient intellectual work to familiarize themselves with "the whole result." Hence, "knowers and producers of knowledge need not be identical." It is unclear whether such a proposal could translate across diverse social scientific examples or into social-epistemological issues where the quasi-Popperian limits of the physical sciences might hold differently, if at all.

Thoughts of this sort are absent from Hardwig's discussions of epistemic dependence and trust: their absence might be read as irrelevant to the social-political effects of his conclusions. Their absence from Tolstoy's example is less benign since the consequences of his inquiry were far-reaching for the people whose lives were investigated. Yet it is not easy to imagine that such stereotype-infused patterns in a highly stratified society could have been eliminated from the investigations, and from their real-world consequences. Behind a veil of ignorance, it might be sufficient for generating trust in the knowledge at issue, in Hardwig's experiment: to assume that the scientists were sufficiently qualified to merit trust. By contrast, in Tolstoy's investigations more is required: juxtaposing the examples suggests a complex contrast. The absence of such considerations from the Tolstoy example shows one salient reason why this absence matters: it urges re-reading the Hardwig example to ask again about his "named" scientists whose epistemic neutrality cannot so readily be assumed, after all.

Clearly, it is impossible to legislate for bias-free, stereotype-free inquiry. While a certain homogeneity of training will have been required of Hardwig's scientists, together with different levels/areas of expertise according to their place in the project, with Tolstoy's investigators, their named differences in rank and social "station" seem to ensure sufficient diversity to allow for epistemic negotiation. Yet, the example conveys a sense of homogeneity within the ranks that makes so open a process unlikely: an epistemology without a knowing subject can make no space for the specificities of epistemic accountability. The difference between the two is not quite so stark, though, for accountability is more narrowly circumscribed in Hardwig's example than its significance signals: still, there is no space for epistemic *care*. If an investigation is to do epistemic justice to the subjects/objects it studies, it has to convey respect for plurality and dissent among the investigators, not because diversity itself is an unqualified good, but neither is consensus derived from presupposed homogeneity, which may be merely a local effect of that sameness. Declaring that the answers were *not* the product of human thought could be a compressed affirmation of differences being negotiated to a point where taking individual or otherwise disparate thoughts into account was no longer necessary: where specificity and diversity had been transcended. The report of the report is not persuasive in this regard.

To think and act well about larger social-political consequences of land reform, the emancipation of the serfs, and the inclusion of these "minorities" within a renewing social order, Tolstoy (as Alexandrovich) needed an investigation that would facilitate "the racial minorities" being well enough known for responsible governance and administrative measures to be initiated. Yet there is no indication that these investigations were open to learning about or from them, in the specificities of their circumstances. Tolstoy is writing of a society undergoing radical social change. Setting aside issues about whether the "minorities" would have a voice, one must also ask whether there was room to *care* about the "minorities"; or whether the outcome was determined before the fact. Given the answers' declared indubitability, there will be minimal space for criticism *after* the fact. When the inquiry is completed, the investigation is closed: it seems, in fact, to have been closed even before it began. While my initial contrast may cast Hardwig's example as benign, even democratic, his investigation, too, is structurally closed in these multiple aspects. His model of cooperative inquiry offers no space to think about uneven social arrangements and about just distributions of public epistemic authority, in everyday circumstances.

In the second section of this chapter, I have considered two enactments of the concept of knowledge in "our social cognitive ecology." Both are self-contained, epistemologically: in both universal, necessary and sufficient conditions for the existence of knowledge are less readily discernible than epistemologists and philosophers of science after logical positivism may have intended. The subject matter of a physics experiment or of a social scientific inquiry assumes a significance in both "contexts"—discovery and justification—that it might not claim in orthodox philosophy of science/theory of knowledge. Despite its explicit situatedness, the Tolstoy example poses questions about its own social-political reach: about its capacity to know the subjects of inquiry well enough to do well by them, with a sensitivity to how they *must* be known if new systems of governance are to accommodate *them* as who they are and how they are situated. A rejoinder might be that nothing should be expected of a fictional example beyond the "world" the novel creates: it is a plausible thought. Yet here I am stretching the example beyond its confines to give a reading that contributes to illustrating some of the central epistemological claims of this chapter. Whether it succeeds remains an open question.

Admittedly, it would be bizarre to expect Hardwig's experiment to manifest all the possibilities mentioned here: his worries about the consequences of acknowledging how much "our" rationality rests on trust may merely be "paper doubts." Furthermore, there is apparently no room for dissent among contributors to the physics experiment once the problem is set and the tasks are distributed: such loss of individual autonomy seems to contribute to, not to detract from, knowing well. Hardwig is writing of a physics experiment where ninety-nine participants were epistemically interdependent throughout the inquiry for results they could not, separately, verify for themselves. He thus produces a strong case for the centrality—the indispensability—of trust in making the experiment's conclusions possible. His analysis, for many social epistemologists, counts as an exemplary contribution to wider understandings of the significance of trust in the production of knowledge. Yet I am revisiting Hardwig's essays to propose that, despite the place they occupy in social epistemology, the analysis does not translate well enough into a model for the role of *trust* in knowledge to inform a robust social-ecological epistemic position. Its explanatory power is too slender for it to count as exemplary for evaluating epistemic practices more generally; or for showing how trustworthiness and credibility, conceived more widely than within an enclosed community of scientists, can be well enough established to promote epistemically-ecologically viable social cognitive relations and practices. A structural limitation is built into his physics example by its restriction to a homogeneous population of equally positioned and accredited knowers, and into the exemplary status it accords to physics as a social-epistemological practice. The implications of his observations are generalized to propose conclusions they do not—perhaps cannot—support, given the esoteric epistemic conditions of the physics experiment. Hardwig's claims for the place of trust in scientific knowledge might be judged so commonplace, in the twenty-first century, as to require no further analysis. Yet a renewed emphasis on testimony in feminist, anti-racist, and other "situated" epistemologies, together with innovative developments in epistemologies of ignorance and analyses of epistemic injustice, afford this work a renewed, if controversial, salience for thinking about the place of knowledge in "our" social cognitive ecology.

The two examples adduced in this section of the chapter are analogous in suggesting some vital implications of conceiving knowledge-making as a collective/collaborative practice: they are enacted differently with respect to the kinds of knowing and to the place and/or situation of the participant knowers within them. They point to diverse understandings of the place of knowledge in relation to human cognitive authority whether in a specific, relatively fixed domain, or across incipient upheavals in a social order. More vitally, they caution against encouraging a collective imagination too narrowly focused on dislocated conceptions of truth, fact, and certainty to the neglect of engaging with the vital epistemological significance of affect and community, together with the conditions and circumstances of their ecological and civic specificities.

Chapter 4

Particularity, Epistemic Responsibility, and the Ecological Imaginary

Against the tacit (if inadvertent) assumptions of human "sameness" that silently shape the discussion in the previous chapters, and with attention to how matters of concern can claim legitimate places in thinking about "knowledge in general," this chapter considers diverse enactments of human subjectivity and epistemic agency within certain demographic, geographic, and cultural locations. Despite its endorsement of particularity as a focus of attention, it presumes neither to offer a comprehensive global analysis nor to require a rigid endorsement of hard-edged categories and kinds. It aims (after Yancy)[1] to focus on a selection of disparate examples and situations en route to considering their pertinence for thinking about credibility and vulnerability in testimonial uptake. The intention is not to adduce these situations as exonerating or enabling factors, but to consider their disparate explanatory powers in some of the diverse circumstances central to "everyday life" in the Western/Northern world, where discrete situational specificities have often to be taken into account, accommodated, across a range of disparate endeavors. While the elusive quality of specificities and of discrete situations pushes against assumptions about universal human sameness, interchangeability, and excessive diversity, radical difference cannot be the only plausible default alternative for engaging with human and

---

1. George Yancy, ed., *The Center Must Not Hold: White Women Philosophers on the Whiteness of Philosophy* (Lanham, Maryland: Lexington Books, 2010).

situational specificities. The background question—and puzzlement—is about how, at once, to think responsibly and well about specificities and differences while remaining vigilant against doing epistemic violence to either, or to both.

Two thoughts animate a focus on particularity as a way into this discussion: they may appear to be antithetical, to pull against one another, but this tension is productive in the engagements it generates. The first is from Australian philosopher Val Plumwood, who observes wryly, but aptly: "Women have represented particularity in contrast to male universality . . . and necessity in contrast to male freedom."[2] This might seem to be a small thought, and in some respects it is, but it is potentially risky to invoke it as a way into thinking about particularity. Doing so could seem to confirm an old adage affirming that women and other Others (from a white patriarchal male norm) should stay away, should avoid what in this negative casting amounts to a *descent* into particularity, after all. A tenacious conviction holds, philosophically and in "everyday life," about the value of a certain aloofness. The thought is that they/we need to be wary of entering the messiness of the concrete, the trivial, the everyday: wary of eschewing attempts to attain such higher levels of reason as are aligned with the universal purity of an abstract, principled detachment that transcends the everyday. Invoking an engagement with particularity as a challenge to the presupposed universal seems to affirm Immanuel Kant's claim that a large proportion of humanity, "including the entire fair sex," will be unable to make unsupervised use of their rational powers.[3] Engaging this tension involves refusing its negative antifeminist, "down among the women" implications, while reaffirming its subversive social-political pedagogical potential. This thought is a guiding thread in this chapter.

The second thought, also cited in chapter one, is from Adriana Cavarero, who observes: "Uniqueness is epistemologically inappropriate." She proposes, provocatively, that in establishing the overarching superiority of a putatively universal *logos*, "the philosophical tradition . . . ignores the unrepeatable singularity of each human being, the embodied uniqueness

---

2. Val Plumwood, "The Politics of Reason: Toward a Feminist Logic," in Rachel Falmagne and Marjorie Hass, eds., *Representing Reason: Feminist Theory and Formal Logic* (Lanham, MD: Rowman & Littlefield, 2002), p. 21.

3. See Genevieve Lloyd, *The Man of Reason: 'Male' and 'Female' in Western Philosophy*, 2nd ed. (London: Routledge, 1994), p. 66.

that distinguishes each one from every other."[4] The remark suggests that in engaging responsibly with the everyday specificities and vagaries of human lives—epistemologically, pedagogically, morally, politically, and socially—"we," whoever we are, have to counter any assumption that *particularity* merits epistemic attention only to the extent that it informs the universal. Endorsing Cavavero's contention and responding to the challenge Plumwood poses, in this chapter I consider some implications of an injunction against uniqueness together with a (quasi-generic) feminization of particularity—as it has infused Anglo-American theories of knowledge, especially in the *moral epistemology* that informs white Western/Northern social and political practices reliant, if tacitly, on knowing people well enough, in their singularity. As I have observed, for Cavarero, even "philosophies that value 'dialogue' and 'communication' remain imprisoned in a linguistic register that ignores the relationality already put in action by the simple reciprocal communication of voices."[5] She cautions against complacently imagining that the anonymous, monological pronouncements of the white Western tradition have been displaced in favor of an emphasis on relationality. Still in 2016, referring to the "autocratic, integrated and cohesive ego" of the Western masculinist tradition, Cavarero proposes *not* fragmenting the subject, but "instead of breaking its vertical axis into multiple pieces, one could try bending it, giving it a different posture . . . inclining the subject toward the *other*—as the relational model allows and, from a geometrical perspective, even encourages."[6] Her "critique of rectitude" offers an opening into such ways of thinking.

Engagement with particularity is a recurrent theme for Cavarero: as I have noted, she addresses it eloquently in *Relating Narratives: Storytelling and Selfhood*, where she writes:

> Philosophers themselves—servants of the universal—are the ones who teach us that the knowledge of Man requires that the particularity of each one, the uniqueness of human existence, be unknowable. Knowledge of the universal, which

---

4. Adriana Cavarero, *For More than One Voice: Toward a Philosophy of Vocal Expression*, trans. Paul A. Kottman (Stanford: Stanford University Press, 2005), p. 9.

5. Cavarero, *For More than One Voice*, p. 16.

6. Adriana Cavarero, *Inclinations: A Critique of Rectitude*, trans. Amanda Minervini and Adam Sitze (Stanford: Stanford University Press, 2016), p. 11.

excludes embodied uniqueness from its epistemology, attains its maximum perfection by presupposing the absence of such a uniqueness. . . . *What* Man is can be known and defined, as Aristotle assures us; *who* Socrates is, instead, eludes the parameters of knowledge as science, it eludes the truth of the *episteme*.[7]

The discussion speaks to a commitment to constructing knowledge that fosters democratic, respectful cohabitation and interaction. Again, as I note in chapter 1, woven together, these threads sustain an engagement with particularity focussed on understanding how "we" can know diversity and "difference" responsibly: on determining whose testimony merits a hearing; who is well positioned to understand or oppose practices of determining validity; who is thwarted in established structures of incomprehension and intransigence. The "who" could be a single knower; it could be a group, silenced, marginalized, enabled or validated. The putative knowledge could be an isolated claim or a complex theoretical apparatus, acknowledged, corroborated—or discredited. Hence, they may slip through the cracks of public acknowledgment and support.

These are apt cautions. Such consequences would be neither politically nor ontologically effective. Hence, advocating an epistemological turn toward particularity invokes significant caveats: its purpose is emphatically not to reclaim the stark *individualism* against which feminist and antiracist moral-political-epistemological critique is frequently, and aptly, directed. Particularity, singularity, and uniqueness are conceptually and ontologically distinct from individualism in their philosophical genealogy and in their significance, nuances, and effects. Thus, a turn toward particularity has to navigate a difficult passage between the stark invisibility/inaudibility often generated by the *logos* as Cavarero construes it and the scattered dissolutions that moving too close to pure particularity entails. Such are the conundrums these thoughts generate. Yet particularity has not been entirely ignored in Anglo-American ethics, political theory, and epistemology. Numerous prefeminist/predifference-sensitive moral judgments come down to evaluating particular actions with/to/on specific people: sensitivity to particularity is not new to feminist, antiracist or

---

7. Adriana Cavarero, *Relating Narratives: Storytelling and Selfhood*, trans. Paul A. Kottman (London and New York: Routledge, 2000), p. 9.

other postcolonial epistemologists, even though its implications may need frequently to be re-evaluated.

The epistemology that has traditionally, if silently, informed Western-Northern moral-political thinking has sought to subsume the particular under general, universal precepts, often appropriately so, in the interest of impartiality. But problematically, the particular, the unique, the "case" that does not fit often disappears in such judgments, for want of sufficiently sensitive taxonomic resources. These disappearances can be troubling, especially in their implications for knowing across differences, and for feminist, antiracist, postcolonial participation in the (traditionally white androcentric) politics of testimony. Failure to fit is rarely a neutral, objective property or event, nor is it always deplorable. Michel Foucault aptly observes with reference to items that can find no place "within the true (*dans le vrai*)," that such "failure" can be a consequence of the limitations of the going *episteme*: of established patterns of emphasis or judgement, themselves too frequently derived from the everyday expectations of affluent, educated, white male lives: thence, too often closed to "external" critical analysis.

While Anglo-American philosophers have engaged extensively with moral particularity, an example I cite in chapter two points toward a different aspect of the issue. In the early 1990s, in "Moral Perception and Particularity,"[8] Lawrence Blum suggests that the specifities of moral situations show how subjective variations in perceptions of salience commonly feed into diverse, perhaps contentious, moral judgments. The claim is not that exceptions prove the rule, but rather that expunging particularity en route to achieving universality can truncate the analytic reach of interpretation and analysis. While orthodox Anglophone neo-positivist epistemologists may urge suppressing particularity for its capacity to obstruct paths to the universal, feminist, postmodern, postcolonial and other (potentially subversive) twentieth- and twenty-first-century epistemologists frequently start from the particularities, the specificities of women's lives and the lives of people otherwise Othered in the hegemonic discourses and practices of the white Anglo-American mainstream. They chart the epistemological significance of knowers' situatedness in specific relations to the circumstances, events, people, and places they seek to

---

8. Lawrence Blum, "Moral Perception and Particularity," *Ethics*, 101, no. 4 (July, 1991), pp. 701–25.

know, and in their relations to/with other knowers. In its hospitality to the once-outrageous question, "Whose knowledge are we talking about?" feminist epistemology prepares the ground for addressing many of the questions Cavarero poses; but there are differences.

Yet in its omissions, Blum's analysis highlights a facet of Cavarero's approach where she parts company with, and speaks a different language from, the usual places where particularity arises in Anglo-American philosophy, both feminist and otherwise. Blum's analysis of moral perception follows a third-personal pattern, showing how X perceives Y in certain particular (Z) circumstances. By contrast, Cavarero's recommendations are addressive, relational, expressed in a language that recalls Annette Baier's and my own arguments for a language of "second persons." Striking is the place Cavarero grants to what she calls "relating narratives" where "relating" refers both to narratives that *forge* relations between and among people, and those that *relate* particular circumstances, feelings, and responses, narratives that convey the "who" that, in her view, disappears into the "what" of third-personal tellings. This is not a minor difference.

Relationality for Cavarero is no mere dyadic, two-voiced replica of the univocal, monologic singularity of impersonal, abstract individualism. Rather, it is from a reciprocal relationality such as she—still following Arendt—advocates that social, political, and *epistemic* community and affiliation are generated, contested, sustained between and among selves who bear only a distant resemblance to the abstract frozen self of modernity. Perhaps enigmatically, Cavarero observes: "[W]omen are usually the ones who tell life-stories . . . like Penelope, they have since ancient times, woven plots with the thread of storytelling. Whether ancient or modern, their art aspires to a wise repudiation of the abstract universal, and follows an everyday practice where the tale is existence, relation and attention."[9] How these thoughts connect with the materiality and concrete urgency of oppression and marginality is a question which such seemingly peaceable narration may be challenged to address, but Cavarero's emphasis on responsibility and mutual exposure opens space for engaging with it, especially when *epistemic* responsibility figures prominently among responsibility's modalities.

Cavarero's thinking about relationality is indebted to her reading of the place of *natality* in Hannah Arendt's thought, in its attentiveness to the exposure of the newly born: an exposure that, for Cavarero, is

---

9. Cavarero, *Relating Narratives*, p. 54.

ongoing, constitutive of human lives from birth to death. It is an exposure many philosophers of modernity, in their press toward achieving autonomy and realizing the self-protecting "buffered self" (in Genevieve Lloyd's apt phrase)[10] have sought to escape and indeed to deny. This "self" fears the mutual exposure and openness of its "thrown-ness" into the world, to borrow a powerful metaphor from Martin Heidegger. Yet people are fundamentally and on-going-ly exposed to one another, and that exposure, either in attempts to paper it over and defend against it, or in philosophies and social-epistemic practices that endeavour to know, understand, protect and honor it, is instrumental in shaping personal, educational, social, political interactions, all the way down.

Some of these thoughts arise out of developments in social epistemology, where testimonial practices—simple and/or complex exchanges between and among people in the real world, ongoing informed contestations—have achieved a new philosophical salience, making way for engaging relationally, with knowledge.[11] Noteworthy in this respect are the seemingly small yet far-reaching effects of linguistic shifts in this inquiry: shifts away from impersonal, monological, polite, third-person propositional variants on the "S knows that p" rubric to a language of speakers and hearers: away from a spectator model of knowing toward an interactive addressive, responsive mode. Edward Craig's *Knowledge and the State of Nature*, Miranda Fricker's *Epistemic Injustice*, and José Medina's *The Epistemology of Resistance* count among pertinent texts in this regard.[12] Albeit diversely, they open spaces for thinking and acting *away from* practices of regarding testimony as comprised merely of information-conveying statements reporting everyday "facts," and toward engaging with putative interlocutors and with the affect-laden

---

10. Genevieve Lloyd, *Providence Lost* (Cambridge, MA: Harvard University Press, 2008), p. 322.

11. C. A. J. Coady, *Testimony: A Philosophical Study* (Oxford: Oxford University Press, 1992); Michael Welbourne, Knowledge (Montreal: McGill-Queen's University Press, 2001), and *The Community of Knowledge* (Aberdeen: Aberdeen University Press, 1986); Jennifer Lackey and Ernest Sosa, eds., *The Epistemology of Testimony* (Oxford: Clarendon, 2006). See also my "Testimony, Advocacy, Ignorance: Thinking Ecologically about Social Knowledge," in Adrian Haddock, Alan Millar and Duncan Pritchard, eds. *Social Epistemology* (Oxford: Oxford University Press, 2010), pp. 29–50.

12. Edward Craig, *Knowledge and the State of Nature: An Essay in Conceptual Synthesis* (Oxford: Clarendon, 1990; Miranda Fricker, *Epistemic Injustice: Power and the Ethics of Knowing* (Oxford: Oxford University Press, 2007); José Medina, *The Epistemology of Resistance* (New York: Oxford University Press, 2013).

world, through testimonial practices broadly conceived and enacted. Epistemic individualism ceases to be a foundational assumption, for testimony functions as *relating* in both of Cavarero's senses and as *exposure*, encompassing vulnerability and requiring trust. Significant is the attention social epistemologists give to specifically situated epistemic negotiations: to elaborated testimony enacted in narrating back and forth; generating relations between teller(s) and hearer(s) that politically—be they benign, malign, or neutral—exceed the one-liners uttered on a putatively level playing field on which standard Anglophone testimony literature often relies. Such practices commonly require extended, collaborative hermeneutical engagement, deliberations back and forth, struggles to understand and respond. Nor will they be dislocated, smooth, or benign. Recalling her discourse-changing introduction of the concept "situated knowledges" into feminist and other epistemological inquiry, in 2016 Donna Haraway affirms: "We relate, know, think, world, and tell stories through and with other stories, worlds, knowledges, thinkings, learnings. . . . Other words for this might be materialism, evolution, ecology, sympoieisis, situated knowledges."[13] Where reliance on the *S-knows-that-p* rubric enabled Anglo-American epistemologists to imagine they could transcend the vicissitudes of the world by establishing universal, a priori, necessary and sufficient conditions for "knowledge in general" Edward Craig, as his reference to the "state of nature" signals, returns to the world as the place where knowledge is made, contested, deliberated, adjudicated. A distinction he draws early in the book is germane to the significance of this shift to an addressive mode, both for the purposes of this inquiry and, I suggest, for epistemological and educational purposes broadly conceived. Recall Craig's observation: "There are informants, and there are sources of information. . . . Roughly, the distinction is between a person's telling me something and my being able to tell something from observation of him."[14] Craig focuses on engaging with people as interlocutors: as informants, where "the *distinctive* relation of believing another person"[15] is pivotal, and where epistemic emphasis is as much

---

13. Donna J. Haraway, *Staying with the Trouble: Making Kin in the Chthulucene* (Durham, NC: Duke University Press, 2016), p. 97.

14. Craig, *Knowledge and the State of Nature*, p. 35.

15. Craig, *Knowledge and the State of Nature*, p. 36. See also Richard Moran, "Getting Told and Being Believed," in Jennifer Lackey and Ernest Sosa, eds., *The Epistemology of Testimony* (Oxford: Clarendon, 2006), pp. 272–306. I have benefited from Moran's analysis in what follows.

on the structure of the engagement as on the knowledge/information exchanged. At issue are the speaker's (and hearer's/hearers') responsibilities in situations both trivial and complex: someone who acts as my informant gives me her *assurance*, tacitly affirming I can rely on what she says. A certain particularity is involved, according to the detail of the situation and the substance of the exchange, for it matters who the other person is: a responsible hearer will not believe just anyone, unless perhaps for quotidian information—about the time of the train, the location of the bank. Hence, responsibility is not the speaker's alone. It is relationally enacted. Not only are giving assurance and taking responsibility interactive, then, but when communicative/disputational discourse displaces the monologic language of autonomy and rights, the very idea of individualistic epistemic self-reliance is exposed as contestable, implausible, and absolute certainty is increasingly elusive (and rightly so). As I have observed, it is in the attention he accords to the informant(s) in a testimonial event that one of the philosophically innovative aspects of Craig's approach is evident: this is no touristic, spectator theory, but one where people acquire knowledge/understanding from and with one another "as subjects with a common purpose, rather than as objects from which services, in this case true belief, can be extracted." By contrast, interacting with people as sources of information involves an objectification not unlike what is involved in knowing the age of a tree by counting its rings (his example). Nor can it, without contest or confirmation, be assumed to be benign. That said, most of Craig's discussion is of third-person and first-person knowledge (cf. p. 68). As I observe in chapter 1, there is no talk of "second persons" in his analysis beyond this reference to a common purpose; hence a flavor of replicability remains, references to informants notwithstanding. In this aspect, his analyses differ sharply from Cavarero's. Nor does his discussion of informants address the *ethics and politics* of testimony, of asymmetrical social-epistemic positionings of speaker and hearer, or of the multiple ways—be they benign or malign—that relating to people as informants can play into their vulnerability or mask their particularity, as readily as it may affirm their credibility. Recall Cavarero's remark that philosophers who value dialogue ignore the relationality of reciprocal communication. Working with people as informants can be as straightforward as Craig proposes only if speaker(s) and listener(s) are situated on a level playing field, with equivalent expectations of claiming a place to speak, to be heard, respected, contested, believed, (mis)understood: where some (minimal) sense of reciprocity prevails.

Beyond simple exchanges of neutral information—if even there—such ease of communication can rarely be taken for granted.

Consider an example that shows attentiveness to particularity at work in a situation that gives some sense of what is at issue in Craig's distinction: it shows how engaging with people as informants—as *who* they are—invokes urgent yet elusive epistemic responsibility requirements. To begin, it is worth affirming that telling a story about a development project in Africa, as I now do, is also fraught with interpretive issues, for levels of ignorance and unknowing, of assent or refusal—are inevitably multiple. They are compounded by the tenacity of an instituted social imaginary that can block possibilities of knowing whereof one speaks. And, they are further compounded in this story I tell, in my voice, which is a retelling of a story told by another, where the voices of those who figure in it are neither audible to me, nor mine to them. Stories tend to be time- and place-bound, as this one is: hence, they present temporally situated "truths." Thus, this story—a 2007 story—may be read differently now, ten and more years after the events: even though retelling it in 2019 is again an interpretive retelling. For just these reasons, such stories highlight the hermeneutic complexity of real-world events, pointing toward broader implications, be they positive or negative, for everyday and educational practices of endeavoring to know people *well enough*, in their specificities and differences: in their temporal-situational groundedness. It need not/should not be a one-off event: reading or listening to such stories can engage the specificities of their time- and place-boundedness while opening space for multiple layers of interpretation, contestation, conversation. Stories do not reduce to nothing in consequence of their specificity and contestability, but rarely can they be judged "sufficient."

Turning to narrative against a background awareness of the complexities of particularity, epistemic responsibility, and testimony, I now tell a piece of a story about an early twenty-first-century UK development project in Katine, in rural northeast Uganda. Initiated in 2007, the Katine project was sponsored by the UK's *Guardian* and *Observer* newspapers and Barclay's Bank. It was conducted by the African Medical and Research Foundation (AMREF) and Farm-Africa. Investigative journalists from both newspapers spent active time there. Eminent among them is Madeleine Bunting, whose book on the German occupation of the Channel Islands in World War II is an exemplary historical investigation, and who was then (in 2007) a columnist with the *Guardian*. These comments suggest, too briefly, why it is reasonable to read her article as offering empirically based testimonial evidence worth taking seriously: I "know" her work well enough to place my trust, albeit cautiously, provisionally, in her credibility and epistemic/testimonial integrity. Her testimony informs my thinking here. These, then, are her words:

> My hunch is that it would take several months of living down one of those long, meandering dirt paths in the bush to begin to understand how a Ugandan woman sees her life. The first thing a westerner doesn't grasp is the scale of Africa; they always have a 4x4 to jump into, which will speed them to Kampala with its hospitals, shops and embassies. For millions of African women, every journey involves hours of walking. Three hours to a council meeting, two hours to visit an antenatal clinic, an hour to visit a friend to borrow a pen, an hour to get a malaria tablet. At least.
>
> The second aspect of rural African poverty which is so hard to grasp is that most village women have very few manufactured belongings. A couple of dresses, a pair of flip-flops, a few mugs and bowls, a sliver of soap. You need to have nothing to know how precious an exercise book is. It's strange how difficult this scarcity is for us to imagine; on the Guardian's Katine website, bloggers urged Katine residents to build their own desks. "It's not difficult, I could teach them in a couple of days," asserted one of these armchair development advisers. But who buys the nails—possibly an eight-hour-round shopping trip—and with whose money? Where do you get the planed wood in a country where wood is an extremely valuable resource? . . . Who transports it to the remote school? Our lives are so conditioned on the availability of what we need that we have no inkling of what it might be like to live with constant unmet need.[16]

These details, these minutiae of the everyday are constitutive of the lives lived there: of *who* these women were as Bunting perceived and engaged with them. Admittedly, in the account one hears her voice, not theirs; hears only her voice, itself temporally and situationally bounded. A reader needs to surmise that the questions she asks, the everyday details she begins to understand, come from her involvement in what Heidegger

---

16. Madeline Bunting, "Two years on, Katine offers much to celebrate—and much to feel frustrated about," *The Guardian* Sunday 1 November, 2009.

might name the *allgemeine Alltäglichkeit* (the "ordinary everydayness") of these women's lives. The details attest to a certain familiarity, a level of relationality that contrasts with the superficial glance of a passing tourist.

Yet not surprisingly, Bunting was neither universally nor unequivocally praised for seeing and naming the materiality of those lives: criticisms abound in the "Comment Is Free" section of the site where the story is posted, as they would in an epistemic climate that eschews the authoritarianism of truth by fiat. She may not see accurately what is before her eyes, she may be seeing through a pair of glasses on her nose (paraphrasing Wittgenstein) that prompts her to observe certain things and occludes others:[17] she may not know whereof she speaks. But at that time, with these caveats, she opened a path toward rendering intelligible some of the minutiae of these lives. The perceptive questions she poses offer a glimpse into how "their" sense of reality differs from "ours": they *challenge the universality of "our" experiences.* Attending to basic and urgent everyday particularities—the cost and effort of obtaining nails, the scarcity of wood, the distances walked in the heat wearing flip-flops—attending to them attentively and imaginatively, allowing them to unsettle the complacency of a distanced social imaginary based on plenty, not scarcity—learning to imagine beyond them to the myriad other things and possibilities that *are not*—can, for the hitherto unthinking, open the way to the "doxastic shock"—to the incredulity commonly integral to projects of recognizing and repairing epistemic injustice. Doxastic shock carries a potential to destabilize a complacently taken-for-granted familiarity presupposed in unquestioned, disengaged assumptions of human sameness and of the uniformity of living conditions. It can expose the systemic and systematic harms such assumptions enact in rendering the unnameable invisible, keeping it beneath notice, ensuring that these matters continue (in Cavarero's words) to "elude . . . the truth of the *episteme*."[18]

In circumstances such as Bunting relates, occasions where people need to know enough to think intelligently, to ask questions that fit this situation, to debate well with friends, colleagues, students, official agencies, and others—and to offer contest or support, whether moral,

---

17. "Where does this idea come from? It is like a pair of glasses on our nose through which we see whatever we look at," Ludwig Wittgenstein, *Philosophical Investigations*, trans. G. E. M. Anscombe (Oxford: Basil Blackwell, 1968), no. 103.

18. Cavarero, *Relating Narratives*, p. 9.

financial or other—testimony/telling is a principal source of knowledge. It is also a fragile source. Few people who need to know the matter at issue will or can be situated to see for themselves; fewer will be able to relate well enough with the Katine women to expect, responsibly, to know them. Thus, questions about where to place belief, how to accord credibility wisely, look for answers to sources whose capacity to provide good enough reasons to believe needs always to be assessed for its plausibility in taking *their* particularity into account. Conceptually productive are the practices and implications of what I have called *the ecological imaginary*. Here, epistemic responsibilities both "individual" and communal are myriad, and delicate: difficult to discern, yet crucial to deliberate and enact. It is through mazes of such jumbled assemblages of what might and might not be so, of whom to believe and why, of how to know what one will never perceive in the way the S of mainstream epistemology so readily did when he (or sometimes she) knew that p, that knowing the particularity, the uniqueness of the unfamiliar and the strange may have to occur.

Such evaluative practices in epistemic judgments are integral to acquiring, exercising, endeavoring to teach appropriate social-cultural epistemic literacy in twenty-first-century societies, where most of what "we" inquirers claim to know about people and places both close to and far from "home" involves elaborated if remote versions of what Bertrand Russell dubbed "knowledge by description." Typical are assertions about the weather, the opening hours of the library, the price of food. For Russell, standard examples of such knowledge are exemplified in "S-knows-that-p" assertions which commonly convey a sense of perceptual accuracy, at least in materially replete societies. Their paradigmatic status derives from postpositivist affirmations of the simple, direct knowability of empirical "givens" in putatively ordinary experiences, where assumptions about human sameness and about a certain uniformity of "the furniture of the world" prevail. A further unspoken assumption is that both S and p are straight-forwardly available as putative knowers and known, and that the knowing is situation-neutral. Such practices are increasingly contestable in postpositivist recognitions of the politics of knowledge and testimony: this case is no exception. Hence, acknowledging the multiple interpretability—and the contestability—of such tellings, I call this a story a narrative, whose tellers are/were implicated in its form, content, interpretation, and circulation. It claims to offer entry into an area that needs to be investigated rather than presenting a collection of established, putatively unchallengeable facts. Stories like these fall somewhere between philosophical anthropology-ethnography and specifically situated literary works, novels. Even though their self-presentation may be more factual than fictional (or the reverse), such tellings can perform a function akin to the effects of certain novels, which (in Gatens's apt words) "seek to challenge, re-form and transform dominant social imaginaries in order that they may become more inclusive, more ethical, and more just."[19] Bunting's story, despite or perhaps because of its time-bound specificity, sits quite well within this frame.

Nonetheless, this story, in its complexity, does not speak for itself, nor can I claim to have read it knowledgeably. To cite a pertinent reading from a different stance: in the *Southern Journal of Philosophy*, Kristie Dotson engages critically with the Spindel conference paper where I discuss these matters.[20] Specifically, she considers and contests the racial implications of the politics of knowledge, apparent in my reading of this event. She adduces powerful evidence of the limitations of narrated "outsider" examples and counterexamples in their attempts to generate response and critique—to advance understanding across differences, while avoiding temptations to superimpose examples onto events where the fit is less than clear. Read together, the papers enact a compelling critical-constructive debate.

In this aptly titled article, "In Search of Tanzania: Are Effective Epistemic Practices Sufficient for Just Epistemic Practices?" Dotson contends: "Effective epistemic practices in the Tanzanian health care example do not necessarily indicate the kinds of just epistemic practices needed to remove epistemic injustice."[21] Significantly and appropriately, she deplores the absence of Tanzanian *voices* in my article—an omission I had at the time failed to recognize and that attests to a lapse in epistemic responsibility on my part. These were deliberative, situated, negotiative practices: features I applaud. But these same features had a

---

19. Moira Gatens, "The 'Disciplined Imagination': Literature as Experimental Philosophy," *Australian Feminist Studies* 22, no. 52 (March 2007), p. 30.

20. Kristie Dotson, "In Search of Tanzania: Are Effective Epistemic Practices Sufficient for Just Epistemic Practices?" *The Southern Journal of Philosophy* 46, no. 1 (2008), pp. 52–64; the paper to which she was responding was later published as "Advocacy, Negotiation, and the Politics of Unknowing," *The Southern Journal of Philosophy* 46, no. 1 (2008), pp. 32–51.

21. Dotson, "In Search of Tanzania," p. 52.

blinkering effect in my neither noticing nor naming the diverse modalities of voice and interaction such situated inquiry demands. Again, the power of example—Dotson's example—is significant. She reads creatively "against the grain" of the narrative and the events I invoke, exposing a deeper layer of epistemic, racially infused assumptions and practices that are germane to knowing responsibly across differences, especially when the "subjects of the inquiry" are present, alive, and cognizant of the investigation's purposes—but not only then. Failure to enlist their voices is an egregious omission: Dotson's charges are apt. Her reading captures a strong sense of the (intertwined) positive and negative power of examples, especially where time and listening are available to enable the very hermeneutic readings I advocate. Such readings are rarer in Anglo-American philosophy than they could and likely should be: Dotson enacts something akin to a thesis/antithesis/(possible) synthesis process, where each step contributes to recognizing—acknowledging—the *epistemic friction* (in José Medina's sense) produced and reproduced in hermeneutic engagements with extended examples and case studies, where the writer does not know well enough whereof she/he speaks. Read literally, such friction rubs the wrong way, interrupts the smooth to-and-fro passage of analysis. It manifests where new or untried/outrageous ideas come into conflict with settled assumptions, fixed points of view, ways of seeing the world. Yet while Medina does not always do so, here I read such friction as positive, especially in relation to Dotson's reading of my reading of the Tanzanian project. She shows how epistemic friction can spark radical rethinking and action, without denying that it can also disrupt, destroy them. Such processes are ongoing.

Notably, from reading-thinking against the grain, it becomes apparent that it is not only the voices of the participants featured in the story that are missing. Recall Craig's contrast between "informants" and "sources of information." Provocatively, convincingly, Dotson maintains "trained interviewers are a manufactured 'third class' of interlocutors who are authorized by their ability to take up the 'necessary' training to become merely useful sources of information within the structure of evidence-collection procedures that require such sources."[22] Cast as mere sources of information, these agents' identity—their "humanity"—comes to be deemed unworthy of recognition. The "personal" specificities of

---

22. Dotson, "In Search of Tanzania," p. 61.

the interviewers and translators who will undoubtedly have accompanied Bunting and her colleagues' investigation of the Katine project are also lost, in both my telling and Bunting's. Their status as "third class" in Dotson's description is meant not merely numerically, but also descriptively-normatively, in a negative sense akin to "third class citizens": structurally, these interviewers count neither as primary nor even as secondary sources of competence, of knowing and/or understanding. They are merely background figures.

Dotson deplores the *instrumentality* enacted in practices reliant on these "individuals"—these interviewers—as epistemic informants, where she finds scant evidence of the "epistemic cooperation," the apt humility and receptivity, the to-and-fro engagement that could attest to the interviewers and the "subjects of inquiry" being respected and valued as knowers—*as who they are*: hence, not as mere information gatherers, place-holders in a larger project: as conduits. Again, she shows, bringing such heuristic interpretations into conversation can be facilitated through narrated counter-examples sufficiently powerful to open ways toward understanding differently, unsettling inadequately informed convictions, acknowledging their lived significance. No outright challenge to epistemic authority occurs in my essay, Dotson notes: it fails to counter the epistemic injustices she perceives in the shifting epistemic positionings of Tanzanian interlocutors. This is the strongest, most telling, charge. I must plead guilty. Yet even despite these omissions, the example and counter-example show something of the power of examples in the conflicting readings of the case: in the almost inevitable ambiguities it carries. They expose successes and failures, thus attesting to a need for contestation—for epistemic friction—in knowing and working to understand complex situations through enlarged, open-ended interpretive-hermeneutic frameworks. They highlight the poverty of too slender ways of claiming to know, and they point toward diverse ways of understanding and/or contesting events, situations, human interactions, and convictions. They urge an engaged practice, sometimes replicable, *mutatis mutandi*, in situations sufficiently analogous to find epistemic resources in the knowings they recount; sufficiently detailed to demonstrate their failings. Yet they caution against too-hasty assumptions of sameness and difference: they take time, care, deliberation.

Because Bunting's engagement with "the particular" prompts diverse critical responses, it serves in another way as an entry into addressing particularity, and into challenging fixed if tacit epistemological assumptions to the effect that only self-contained, seemingly "finished" events merit consideration and/or can occupy exemplary positions in epistemological analyses. On the refocused understanding I am urging, epistemic inquiry would and should incorporate a larger dialogical-disputational hermeneutical component than dislocated, decontextualized Anglo-American analyses have commonly done: both the situation and the "meta-situation" call out for sensitive engagement, interactions, contestatory-collaborative readings. There is, and should be, a tentativeness, a certain temerity in telling and hearing such stories—they make "us" think. If we think well, carefully, they could generate some degree of the relationality Cavarero seeks, whose absence she deplores: a stretching of the imagination, opening paths toward thoughtful action, informed contestation, sensitive engagement.

An analogous example is Shannon Sullivan's landmark 2007 essay "White Ignorance and Colonial Oppression: Or, Why I Know so Little About Puerto Rico."[23] Posing the questions *to herself* about why she evidently knows so little prompts Sullivan—searchingly, fundamentally, hermeneutically—to reengage with the situations of her life in affluent white America. It urges her to rethink her hitherto settled thoughts about the places where she has learned and not learned about Puerto Rico: to analyze how the assumptions and stereotypes held in place by living as she does have contributed to her knowing "so little." She works to understand how the effects of the ignorance/knowledge that has, ironically for her, passed for knowing has fallen beneath her notice. Sullivan's interest in the essay (sustained with variations in ongoing investigations of "whiteness" in her subsequent thinking) is in how the then-going US public educational system, with its discourses of ignorance/knowledge, had played into the successful colonization of Puerto Rico, constructing Puerto Ricans "as similar enough to white citizens to be capable of Americanization."[24] Hence, Sullivan recognizes that the "epistemic relationship" between Puerto Ricans and Americans living in the United States is "more complex than the simple opposition between ignorance and knowledge indicates." In her view, the ignorance/knowledge thought

---

23. Shannon Sullivan, "White Ignorance and Colonial Oppression: Or, Why Do I Know So Little About Puerto Rico," in Shannon Sullivan and Nancy Tuana, eds., *Race and Epistemologies of Ignorance* (Albany: SUNY Press, 2007), pp. 153–72.

24. All quotations in this paragraph are from Sullivan, "White Ignorance and Colonial Oppression," pp. 154–55.

is effective and appropriately humbling in how it "denies, or at least places under suspicion the purported self-mastery and self-transparency of knowledge." It helps us "to peek behind knowledge of Puerto Rico to see what unknowledges help compose it": unknowings on which it relies for perpetuating the policies and practices that sustain "Porto Rico's" colonial status. This putatively 'local' situation is replicated *mutatis mutandi* across the Western-Northern world, and beyond.

The urgency of this call is apparent, also, in Bunting's Katine story. There, peeling away—dislodging—colonialist unknowings to allow an "armchair observer" to superimpose her/his "knowledge-in-general" of what *she/he* would have done in that (a similar) place, is unsettling. It is especially so when the "knowledge" is superimposed onto a situation about whose specificities he/she can at best claim ignorance-knowledge, filtered through a social imaginary conditioned by ever-present availability and a systemic incapacity to conceive of or engage with the day-to-day materiality—the physicality—of genuine scarcity. While the places/circumstances of the two discussions differ, the example exposes Western imaginaries' inadequacies in their/our attempts to capture the experiences (at least purportedly) of non-Westerners: their attempts to *know* across radical differences. Hence, although the Katine project and the situation of Puerto Rico relative to the United States are so remotely analogous that to draw them together for purposes of comparison risks doing epistemic violence to each, the issues overlap and merit further investigation, especially with respect to "outsider readings." These issues are central to the discussion that follows.

To consider these matters more closely, I turn to the third part of the title of this chapter: the ecological imaginary. It is through mazes of such jumbled assemblages of what might and what might not be so, of whom to believe and why, of how to know what one will never perceive in the way the 'S' of mainstream epistemology so readily did when he (or sometimes she) knew that *p*, that knowing the particularity, the uniqueness of the unfamiliar and the strange may have to occur. Here, epistemic responsibilities both "individual" and communal are myriad, and delicate: difficult to discern, yet crucial to deliberate and enact.

I have noted that the hegemonic instituted social imaginary of the affluent white Western-Northern world is one of taken-for-granted availability and access: of ways of life where individual self-reliance is an overarching virtue, and where the illusion prevails that such virtue is its own reward. It is easily sustained in widespread convictions (again, in affluent white societies) that "most people," leaving the extension of the term vague, are positioned to achieve their "goals" and fulfil their "needs" by their own efforts. Such needs, be they material or social, are imagined to be natural, the *sine qua non* of a viable human life; and scarcity, when it occurs, is cast as temporary, aberrant, contingent; and sometimes blameworthy. It can and should be "fixed," resolved. These thoughts attest to what, in *Ecological Thinking* and elsewhere, I call an imaginary of mastery and control; an imaginary for which ethical self-mastery, political mastery over unruly and aberrant Others, and epistemic mastery over the "external" world pose as readily attainable goals for the inhabitants of a post-Enlightenment white Western-Northern world. These discourses of "mastery" (the term is Plumwood's[25]) are so closely woven into the fabric of the social imaginary they sustain as to be virtually imperceptible except in their absence—and unchallengeable. They derive from and underwrite reductive ways of thinking for which epistemic and moral agents are isolated units on an indifferent landscape, to which their relation, too, can be one of disengaged indifference. They enlist ready-made, easily applied categories to contain the personal, social, physical-natural world within a neatly manageable array of "kinds," obliterating particularities and differences in order to assemble the confusion of the world into maximally homogeneous units. The human subject—the moral, political, epistemic agent who populates the thinking and the social order informed by this imaginary—is the Enlightenment and post-Enlightenment "man of reason": white, propertied, educated, materially, affectively, and rationally self-sufficient, unified within and transparent to himself. He continues to stand as the presumptive character ideal—and sometime caricature—of white prosperous Western/Northern societies.

In *Ecological Thinking*, one of my principal aims is to show that an ecological remapping of the epistemic terrain can generate transformations in the social order (potentially) capable of destabilizing the epistemologies of mastery, unsettling their hegemony. These many years later, that claim may seem merely fanciful, empirically implausible. Yet such remappings will inevitably be gradual. At their best, they maintain a wariness of master narratives, premature closure, reductionism: they may thus appear to restrict the range of justifiable, definitive knowledge

---

25. Val Plumwood, *Environmental Culture: The Ecological Crisis of Reason* (London: Routledge, 2001).

(and often they do). They urge vigilance in discerning and contesting irresponsible, careless, too-swift knowings that fail to do justice to their objects of study. Ecological thinking is not simply *about* ecology or *about* "the environment": it informs and animates a renewed and renewing *instituting* social imaginary, an ecological imaginary, which endeavors to unsettle and dislodge the starkly individualistic imaginary of mastery and control, to concentrate on *connections*—negative or positive—and relationality. Thinking ecologically is a thoughtful practice: it carries a responsibility to know somehow more *carefully*, more *deeply*, more *cautiously*—yet often more courageously, more boldly—than single surface readings commonly do. As I have observed, it is about imagining, crafting, articulating, and endeavoring to enact principles of ideal (human and other than human) cohabitation. Thus, the ecological subject differs markedly from autonomous man, whose assumptions of mastery over all he surveys allow surveying to substitute for engaged participation, and mastery to suppress acknowledgments of diversity, all in the interests of instrumental simplicity. Ecological thinking offers a conceptual framing for responsive-responsible theories of knowledge and subjectivity: responsive to particularity, diversity and community; responsibly yet boldly committed to knowing well and to countering the oppressions and exclusions the epistemologies of mastery sustain. Rather than aiming to dissolve epistemic friction, it engages and deliberates within it. In these and other ways, it informs my thinking about particularity.

As Cavarero and Sullivan variously affirm, such deliberations about particularity tend to be blocked in white Anglophone Western philosophy and everyday thought and action by an instituted social-epistemic imaginary through which people in affluent Western/Northern societies learn to see and know the Other: the foreign, the distant other; and likewise the other near at hand. Once recognized, such blockages generate obligations to ourselves and to our students, colleagues, and everyday interlocutors to interrogate the fixity, the intransigence of this imaginary in ways that problematize and may destabilize it. Yet undoing the imaginary of mastery with its embedded assumptions of human sameness and replicability will not be achieved with isolated idiosyncratic counter-examples, readily dismissed as mere aberrations and again discursively contained. It calls for gradual recognitions, challenges, disruptions of a kind Castoriadis initiates when, to the instituted imaginary, he opposes the *instituting* imaginary. Instituting processes are generated cumulatively, sometimes boldly, often haltingly to and fro, in the sustained critical-creative activities of a society (or part thereof) exhibiting its integrity in its capacity to put itself in question: in the ability of some of its members to act, collectively, from recognitions that the society is incongruous with itself, with scant reason for self-satisfaction. Imaginatively initiated counterpossibilities, practices of refusal, interrogate the social structure to destabilize its pretensions to "naturalness" and "wholeness," to initiate new makings.[26] Feminist and other postcolonial epistemologists are engaged in creative reconstructions, recognitions, disruptions, such as these.

Implicit in these thoughts is a proposal that narratives—relating narratives—can play a central part in highlighting incongruities that could contribute to destabilizing an established social-epistemic imaginary. That said, it would be overly simplistic to propose that narratives *tout court* offer solutions to the problems that puzzle us, for no narrative can tell "the whole story," even if the idea of a whole story were remotely intelligible. Yet, revisiting Gatens's proposal that I cite earlier: novels and stories—including the one I have told, following Bunting—"seek to challenge, re-form and transform dominant social imaginaries in order that they may become more inclusive, more ethical, and more just." Viewed as "creative cultural criticism," still citing Gatens,[27] they can participate in enabling people, who are otherwise unable or unwilling, to think our/their way out of an entrenched imaginary, where educating our imaginations, and the imaginations of our students, children, colleagues, friends, and other interlocutors is a vital task facing those of "us" who occupy positions of privilege and trust, as educators—or just as "people."

Still, complexities and hesitations abound: this proposal is not that a story, whether factual or fictional, should be taken at face value, nor is it always reasonable to read stories—even first-person narratives—as unequivocally "true." There are many ways to read stories: literally, as "offering just the facts"; cautiously, recognizing that no single narrative can be expected to do that; phenomenologically, for its part in stripping away, in "bracketing" assumptions about how well "we" know/understand/imagine circumstances so radically unlike ours that they only with difficulty find a place—an uneasy place—within the imaginary of having, of plenty, and of availability; and heuristically as initiating "doxastic shock." Knowing

---

26. Cornelius Castoriadis, "Radical Imagination and the Social Instituting Imaginary," in Gillian Robinson and John Rundell, eds., *Rethinking Imagination: Culture and Creativity* (London: Routledge, 1994), pp. 136–54.
27. Gatens, "The 'Disciplined Imagination,'" pp. 29–30.

from stories often calls for approaching them "blank": for a suspension of belief and disbelief. Such knowing can initiate productive critical inquiry into taken-for-granted, seemingly matter-of-course assumptions. That said, even people (singular or plural) approached openly, respectfully, as informants can be unreliable narrators, ready rather to manipulate the truth than to tell it straight. Nor will engaging with rogue particulars regularly destabilize complacent assumptions in the same ways; yet working by analogy offers lessons to be learned. Participants may come away with ignorance/knowledge, in Sullivan's provocative sense; yet as she herself shows, ignorance/knowledge, too, can open places for destabilizing an entrenched social-epistemic imaginary.

Kristin Shrader-Frechette contributes to countering ignorance in the balance she achieves between objective scientific evidence and detailed ecological-environmental harms, with their ethical-political-ontological implications for human sufferings.[28] Her thinking facilitates extending this inquiry into engagements with the politics of care in techno-science. As I have noted, social epistemology's conceptual shift away from uttering knowledge claims as though into a void and toward the language of speakers and hearers, where credibility and trust are central, is germane to the pedagogical and broader significance of these observations. Shrader-Frechette exposes the importance of individual-social-communal uptake for testimony's being heard, understood, contested, acted upon, in communities where power-privilege differences enhance or inhibit a fair, just hearing. Testimony, with its reliance on experts and putative experts is vital to ecological thinking, since nonexperts often have little choice but to trust the "experts," and "individual" epistemic self-reliance is merely a (bad?) dream. Still, the politics of testimony are fraught, drawing ecological-environmental issues into a frame where matters of *particularity* occupy a central place: they prompt questions such as "How can we know well enough to respond well to the effects of the current crisis for *these*—specific—issues?" They are challenging, for standard theories of empirical knowledge focused on "knowledge in general" tend to have little patience with particulars, specificities: the risk of descending into "particularism" haunts them. Social epistemology's reliance on testimony makes space for such questions, showing how well-narrated analyses can illuminate and/or contest complex social-political matters.

---

28. Kristin Shrader-Frechette, *Environmental Justice: Creating Equality, Reclaiming Democracy* (New York: Oxford University Press, 2002).

It exposes the uneven effects of "manufactured uncertainty" for diverse populations, especially those not represented in the figure of "autonomous man": for reclaiming advocacy as a responsible epistemic practice, for its attention to ignorance and skepticism, and its pertinence to educational theory and practice.

Epistemology and the Social Imaginary

> I believe there is no intellectual activity that is not grounded in an imaginary.
>
> —Michèle Le Dœuff[29]

For social epistemologists in the early twenty-first century, newly insistent questions about the ethics and politics of knowing and the social-political positioning of putative knowers struggled to claim a voice in a reimagined epistemology. Like women whose experiences of sexual harassment found no (hermeneutic) place in philosophical discourses of harm and injustice before a sufficiently sensitive conceptual apparatus was crafted to address them, philosophers who sought a place in epistemology-ethics for questions about epistemic identity could find no space to address them in the postpositivist discourse of Anglo-American philosophy. There seemed to be no room to articulate the issues or elicit the acknowledgment (in Wittgenstein's sense) for questions and contestations about epistemic identity and the specificity-singularity of "knowers." A Popperian desideratum—the dream—of developing an "epistemology without a knowing subject"—persisted as an overarching goal.[30]

Posing a question consonant with those Linda Alcoff poses in "Epistemic Identities," I wonder what the consequences might be were epistemic inquiry to focus on the intricacies of the social *imaginary*, rather than on positivism-derived standards of objectivity and proof. The social *imaginary* carries an impressive explanatory-interpretive potential,

---

29. Michèle Le Dœuff, *The Sex of Knowing*, trans. Kathryn Hamer and Lorraine Code (New York: Routledge, 2003), p. xvi.

30. Karl Popper, "Epistemology without a Knowing Subject," in his *Objective Knowledge: An Evolutionary Approach*, rev. ed. (New York: Oxford University Press, 1979), pp. 106–52.

especially in a conceptual framing where power figures prominently and likewise, in analyses that eschew epistemic individualism. Such power—the power of the social imaginary—is socially diffused through practices of knowing, thinking, acting both "individual" and collective, in twenty-first-century philosophy and in the wider world. Just as the tacit individualism of orthodox Anglophone empiricist epistemology truncates inquiry's epistemic promise, the ontological assumptions it carries are troubling, and not superficially so.[31] These issues figure prominently and productively in Cornelius Castoriadis's thinking, and in Michel Foucault's. Yet even as the book's tacit individualism (which is apparent in its references to "the imagination") truncates the inquiry's epistemic promise, the ontological assumptions it tacitly carries are troubling and not superficially so. This is not merely a verbal quibble. In a conceptual framing where "power" figures prominently, if silently, there is greater explanatory potential in the social *imaginary* than is available from enlisting "the social imagination."

Consider how hermeneutical and testimonial injustice are reciprocally constitutive in their identity-constructive/destructive effects. They operate differently one from the other in that testimonial injustice tends to manifest as individual-to-individual harm, even in its reliance on features of social-group identity. Hermeneutical injustice, with its roots in a collective interpretive-conceptual resource, invokes different if analogous culpability issues, deriving from institutional and/or social policies and practices held in place by what—following Castoriadis and Le Dœuff—I call an instituted social imaginary, and not for the words alone. Thus construed, the "imaginary" carries the phenomenologically and psychologically-socially sensitive resources to engage with complex social phenomena that are everyone's and no one's. The conceptual framing that tacitly informs thinking shaped by *the social imaginary* has much to recommend it with regard to understanding, and working from, these issues. Its explanatory potential and political efficacy are noteworthy. Likewise, the language of epistemic injustice analyses is more persuasive singly than collectively, even though both epistemic and ethical subjects risk appearing as more autonomously self-contained than is plausible within

---

31. See Linda Alcoff, "Epistemic Identities," Episteme 7, no. 2 (2010), pp. 128–37; and Elizabeth Anderson, "Epistemic Justice as a Virtue of Social Institutions," *Social Epistemology* 26, no. 2 (2012), pp. 163–73.

the scope of these deliberations. The ethical task is more challenging than any claim for its primacy suggests, while its implications can be deeply political without ethics reducing to politics. Again, this analysis is as political as it is ethical in pointing toward the *systemic* operations of power as they manifest throughout entrenched social imaginings. An epistemic injustice analysis (contrasted with hermeneutical injustice) is more persuasive singly than collectively.

Because no social imaginary is seamless, in the gaps—the interstices—of its interpretive reach, there is room for dissent when/if a wave of justice-motivated collective refusal and creative renewal is set in motion. Hence, I am reading this conceptual framing as political and ethical at once, especially in its focus on systemic operations of power within entrenched social imaginings. Yet it would be implausible, conceptually and politically, to propose that ethics routinely yield to political expediency or to pressures of vested interest, given the infamously negative effects of compromising ethics in the interests of political gain. These are such intertwined, reciprocally constitutive fields that articulating their interactions is more challenging than philosophical claims for the primacy of the ethical suggest. Nonetheless, and significantly, both the ethical and the political (tacitly) presuppose a human-social ontology of self-contained, self-sufficient *individualism* that silently informs the substance of the arguments. Thinking about the wider social/collective resources that need to be considered if philosophers—and others—are to understand epistemic injustice well prompts the question "*whose* everyday expectations are assumed or confounded" in these analyses? Given that such examples are explicitly and diversely populated, the worry may sound bizarre. But any analysis aiming to repair epistemic injustice has to be *systemic*, fully engaged with the social-structural conditions and the specificities that have made injustices possible and that continue to be hospitable to their persisting.

An analysis of the sources of epistemic injustice, therefore, cannot productively be conducted within the purview of settled aspects of more or less homogeneous white affluent lives. Yet for all their innovative intent, some injustice analyses take as their implicit, exemplary counterpoint certain aspects of well-ordered societies whose fabric is torn, often irreparably, by the "injustices" they detail. José Medina captures the problems with working from such assumptions, noting for example: "People's participation in cognitive activities is often impaired by epistemic distortions and biases that limit their capacity to learn, to teach, and to engage in joint epistemic projects."[32] In presupposing a homogeneous society where injustices are aberrations in an otherwise benign social order that calls out merely to be "fixed," there is a certain narrowness of engagement with power that avoids being troubled by epistemic *friction*, in Medina's evocatively productive sense. Following a turn toward social epistemology in the twenty-first century, it is clear that the taken-for-granted goods integral to such analyses need to be exposed and engaged, deconstructed, in a sense contiguous with if not identical to the sense that tacitly informs current analyses of hermeneutic injustice. Without displacing the assumptions commonly integral to white upper-middle class analyses, including their loyalty to a stable ontological-epistemological norm where epistemic injustices are interruptions, disturbances, of an otherwise calm surface and benign social order, efforts to understand the roots of epistemic injustice cannot depart from ethical-epistemological "business as usual" as radically as they might intend.[33]

By contrast, Michèle Le Dœuff enlists "the imaginary" effectively yet not psychoanalytically, to display the richness of its nuances and conceptual resources—although I am suggesting, contrary to her sequential ordering, that hermeneutic injustice is often the primary, more complex and powerful concept from which testimonial injustice derives. Embedded in and informed by collective social interpretive-conceptual resources, the imaginary exposes incidental and systemic epistemic injustices to invoke the culpability assignments held in place by an instituted social imaginary. In consequence, I suggest, hermeneutic injustice comes across (perhaps tacitly) as being temporally and conceptually prior to, and productive of, testimonial injustice in its multiple manifestations. It generates and maintains spaces where testimony can be heard, deliberated, interpreted. Given the extent to which the political and the ethical are intertwined—often positively, supportively—in twenty-first-century Western-Northern societies, it is unclear why their reciprocal, intra-active effects would invite condemnation, as often they do. The very idea of epistemic injustice carries an emancipatory potential that is too frequently

---

32. José Medina, *The Epistemology of Resistance: Gender and Racial Oppression, Epistemic Injustice, and Resistant Imaginations* (New York: Oxford University Press, 2013), p. 28.

33. For ideas in the ensuing pages, I am indebted to a helpful exchange with Gaile Pohlhaus.

absent from putatively neutral liberal discourse, too rarely analyzed and addressed. Hence, the ensuing analysis is at once political and ethical in its focus on systemic operations of power in entrenched social imaginings. In consequence, the power invoked in the book comes across as individually "owned" and enacted, not—more plausibly—as diffused through social structures, as I will propose they must be. The stories are presented as aberrations in an otherwise normalized ethical space. But whose normality is it?

For many philosophers, in such exchanges neutrality would be the taken-for-granted goal. Hence, I read Le Dœuff as tacitly declaring the inadequacy of so putatively elastic a conceptual apparatus for addressing/countering larger social-political injustices in the wider world; for engaging multifaceted iterations of epistemic injustice with its (perhaps silent, yet salient) departure from neutrality. The analysis concludes just as the reader anticipates the arguments' being extended to engage their social-political implications and effects. While it is not clear how, or to what extent, these ideas could move into the political realm, their absence short-circuits the potential of the discussion. In everyday allusions to the wider world, it seems, "wider" rarely means neutral: it seems more commonly to be explicitly structural, multiply interpretable. There are situations where *individual* testimony plays a vital part, yet where injustices are wide-ranging, and as political as they are ethical. Ordinarily, in a settled society, individual testimony speaks and is heard within the purview of such discursive possibilities and constraints. In this regard, the concept-in-practice is much less elastic than it was evidently intended to be, a reading that steps back from its social-political purpose in bespeaking so individualistic a process.

The imaginary, I suggest, has greater explanatory potential than "the imagination" as it appears in the work of Miranda Fricker (and in common parlance); its political efficacy is clearer in consequence of its departure from individualism. Hence, for example, an analysis focused on epistemic injustice can be more persuasive singly than collectively, despite both epistemic and ethical subjects being tacitly depicted as autonomously self-contained. And because no social imaginary is seamless, in the gaps—the interstices of its interpretive reach—there is room for dissent when/if a wave of justice-motivated collective refusal and/ or creative renewal is set in motion. Such analyses are as political as they are ethical, especially when they focus on systemic operations of power within entrenched social imaginings. Still, it would be implausible to contest a reversed ordering outright—to propose that ethics might routinely yield to political expediency or to pressures of vested interest—given the infamously negative effects of compromising ethics in the interests of political gain. These are intertwined, reciprocally constitutive conceptual fields: articulating their interactions is more challenging than claims for the primacy of the ethical suggest. Significantly, both the ethical and the political presuppose a human-social ontology that silently informs the substance of the arguments. Hence, thinking about social/collective resources prompts the question: "*whose* everyday expectations are assumed or confounded in the examples invoked in this early book (and in its revisiting the Stephen Lawrence case)?" Again, since the examples are explicitly and diversely populated, this worry may seem bizarre. Yet in such analyses, some focus on their *systemic* implications is to be expected of they are to be as fully engaged as they could be—to paraphrase Michel Foucault—with the social-structural conditions that have made them possible.

A central problem with analyses that presuppose a settled, privileged social order where power is conceived as individually owned and enacted yet not diffused through social structures, is that the guiding assumption might be that the racism and other injustices[34] apparent in this society are enacted by *people* who are unjust and who just need to "see the light" (as Atticus Finch sees it): a mildly exonerating thought. Problematically, such a claim could "whitewash" the situation: "racism" could indeed be endemic to this society, and yet the focus is on just one example of racial injustice. We are to understand that "racism exists here" while focusing on a single case of racist injustice (and with significant personalizing attention to white characters, Scout and Atticus). Hence, it is vital to acknowledge the tension—hopefully a productive tension—between specificity and generality. The problem remains that excessive focus on the particular can hide the larger framing, the social pervasiveness of racism. At the same time, such a focus on specificity is worthy, not just because this case of racial injustice is worthy of notice, but also because it forces "us" to recognize that this example of racial injustice is itself an example "as conceived by a white female author," presumably for a white audience. *Who* is experiencing racism, and who is describing racism are both significant features of the situation. The

---

34. Racism and sexism are forms of blatant epistemic injustice, of damaging ways of "knowing" people.

difficulty is to characterize such significance in terms that fit within the boundaries of a Western philosophical orthodoxy that routinely calls for readers/thinkers to "leave subjectivity out."

A common rejoinder might once have been that "most" Anglo-American ethics and epistemology would acknowledge no need to articulate such specificities. The analyses and principles they endorse and/or critique are formal and universal: they pertain impartially across moral-epistemic agency throughout the societies where these agents know and act. They function primarily as placeholders in a quasi-formal discourse. Thus, for example, despite Miranda Fricker's assertion (responding to Linda Alcoff) that "neutrality must be the aim,"[35] such identity-neutrality is implausible in view of the—laudable—specificity of the examples that shape arguments intended to inform the real-world consequences and effects of human action: unless the pull of the analytic/postpositivist imaginary is so strong that any departure from neutrality merits reproach. Nor is it apparent *why* neutrality *would be* the aim, in many and diverse social-political-ontological situations where the specific events at issue carry a certain urgency, and a committed response in both words and action is evidently called for. Thus, I am revisiting these creative moments in concert with thinking critically about "the social imagination" and about "epistemic identities," to consider the reach of these conceptual resources for addressing and countering large social-political injustices in the wider world. By "wider," here, I mean less directly "personal" or "individual," more explicitly structural, than situations where *individual* testimony plays a vital part, yet injustices are structurally wide-ranging, and as political as they are ethical. Individual testimony speaks and is heard within the purview of such discursive climates.

Pertinent is José Medina's observation: "The mistake of intellectualism is to think that by changing the epistemic, the ethical and political will follow, whereas . . . people's concepts and cognitions may not control all their emotions, moral characters, and political attitudes." The social *imaginary* makes space for such thoughts. For Medina, and convincingly, "this contextual approach has to be pluralized and rendered relational in more complex ways." In his view, "As fair communicators . . . participants in these exchanges have said, the 'approach needs to be further pluralized and [its] assumptions . . . about the pervasiveness of herme-

---

35. Miranda Fricker, *Epistemic Injustice: Power and the Ethics of Knowing* (Oxford: Oxford University Press, 2006), p. 167.

neutical lacunas and their influence on entire collectivities have to be interrogated.'"[36] But *whose* collective understanding? *Whose* "collective hermeneutical resource" is at issue? Such questions would be anathema to empirical orthodoxy in many philosophical settings; but in socially politically situated epistemic inquiry their omission risks truncating the investigation, dulling its effects.

Among issues central to these thoughts is a need to discern how certain "epistemic identities" and social-structural positionings enable and constrain projects of bringing *epistemic responsibility* into conversation with hermeneutic and testimonial (in)justice. These modalities of injustice are interconnected in the harms they perform, especially in their identity-constructive power, and despite their operating quite differently one from the other. Testimonial injustice is more commonly construed as an individual-to-individual harm, albeit reliant on features of social-group identity; while hermeneutical injustice, with its roots in a collective hermeneutic resource, invokes different culpability issues. Many of these iterations refer to institutional/structural policies and practices held in place, and tacitly enacted, by what I call an instituted social imaginary. For Castoriadis, one of the virtues/strengths of this conceptual choice is in the spaces it creates for thinking, critically and constructively, about a society that is *incongruous with itself*: it could be ethically, epistemologically, politically reprehensible to struggle to preserve an outworn imaginary.

Situating these thoughts within a sedimented social-epistemic imaginary, thus calling for the genealogical analyses many socially constitutive presuppositions require, prepares the way for locating *acknowledgment* among the conditions of its possibility. Ordinarily there is little consideration in mainstream Anglophone empirical inquiry of the situatedness of knowledges (in Donna Haraway's sense) or of the effects of situation—of demography and place—in exposing allegedly neutral power-knowledge structures on which multiple ontological presuppositions of current ecological thinking rely. Admittedly, imagination can be invoked to explicate-interpret many and diverse situations, and with some success: but it is also worth recalling that, in everyday parlance, imagination is commonly aligned with fantasy, untruth, thought experiments, and contrasted unfavorably with knowledge. (Given the book's focus on the

---

36. Medina, *The Epistemology of Resistance*, p. 90.

quotidian, the everyday, this alignment poses a further question: a more basic but nonetheless salient point.)

A viable framework for situating such deliberations is anticipated in questions such as those I pose in 1987, in *Epistemic Responsibility*, despite there being no available conceptual space at that time—no collectively "prepared mind"—for these thoughts to gain traction. Significant conceptual innovations had to occur for them to claim viable discursive space and pertinence. Three factors, at least, are germane to their current plausibility: the growth of social epistemology; a concomitant re-evaluation of testimony as a source of knowledge; and a shift (often attributed to Edward Craig) from an epistemology of monological pronouncements to the language of speakers and hearers. Creating space for such deliberations, these shifts contribute to reweaving the fabric of epistemic discourse in social philosophy.

In Miranda Fricker's early work, stereotypes and the practices they infuse are no one's and everyone's in what she calls a "social imagination": a set of received views and expectations, where members of a society grant them "cognitive sanctuary," to borrow her phrase. In her view, a collective social imagination can be both "an ethical and epistemic liability" and also a "mighty resource for social change."[37] Connections with social power and collectivity appear to be the framing discourse of the argument: they endeavor to establish a "'place'" where creative, subversive challenges to stereotypes and other prejudices gather transformative momentum (the new social movements of the 1960s and after, with the consciousness-raising that animated them, initiated such ongoing challenges). This would be a "place" that could harbor and sustain a certain inertia, a collective complacency about the status quo, a stubborn insistence on the rightness—even despite ourselves—of stereotype-confirming injustices. It is as components of a socially saturating belief structure that many such injustices are best understood: an analysis of individually owned biases curable by isolated moments of empirical counter-evidence cannot account well for their intransigence and elasticity. In consequence, again I suggest, the *social imaginary* offers a viable, capacious resource for explaining systemic-structural operations of belief systems, biases, metaphorics, and imagery. It carries notable explanatory and revisionary potential: it is thoroughly systemic in a quasi-Foucauldian

---

37. Fricker, *Epistemic Injustice*, p. 38, 38n9.

and/or Le Dœuffian sense, and it is explicitly power-infused and politically oriented, hence not constrained by residues of individualism. Therefore, rather than holding ethics and politics apart, I am reading them as co-constitutive—if often healthily conflictual—in generating productive frictions such as those the "social imaginary" captures. The issue is not merely terminological: it is about *framing*, where highlighting certain aspects of these contentious practices renders other aspects invisible/illegible. These aims are intertangled: the imaginary infuses and animates "identity prejudice," while certain putative identities are valorized or denigrated in consequence The curious issue, then, is about how ways of framing these matters highlight certain aspects while making others illegible: how the *framing* (not just the "words") makes legible some of the effects I am addressing, while occluding others.

In the everyday philosophical-political discourse of the late twentieth and early twenty-first centuries, power is conventionally "owned" and individually enacted, and the conception of power that informs these articulations is more tacit than overt. By contrast, Le Dœuff and Foucault position power—plausibly, I contend—as diffused throughout social structures. Invoking the social imaginary opens these issues to genealogical analyses where, for example, an invocation of Foucault on power can sit easily and claim viable explanatory scope. The issues are about the explanatory power of this conceptual apparatus, about how far it can extend in establishing its status as an innovative social-political apparatus. Consider the difference it can make when an explanatory framework derives substantively from the social imaginary as it is articulated by Cornelius Castoriadis, and with equivalent promise by Le Dœuff.

Many of the issues pertinent here revolve around responsibilities to know; most presuppose a social-epistemic "ontology" where being an X carries *epistemic* responsibilities that being a Y does not. Difficult to articulate are injustices performed against hard-won public reasons that generate a certain credibility in favor of scientific research stereotypically presumed to be "neutral" (in a now superseded imaginary). They may unsettle fixed assumptions that accord "science has proved" assertions quasi-inviolate standing, without asking "whose science?" (with a nod to Sandra Harding). Hence, an instituted imaginary of complacency and comfort seeks reassurance from its own persistence, so long as practitioners look away from empirical events that, increasingly, strike at the core of erstwhile certainties: events of denial and refusal which the social imaginary of the affluent Western world may struggle to accommodate, or to discredit.

Especially relevant are Castoriadis's thoughts about the social unrest that follows when a society (or members thereof) where a certain imaginary has prevailed realizes that it is *incongruous with itself*, that it could be ethically, epistemologically, politically reprehensible to struggle to preserve an outworn imaginary. So, epistemic injustices in climate change inquiry are commonly enacted against/upon citizens at the forefront of bringing these matters to public-social attention: matters challenging to self-presentations that overinflate an epistemic exaggeration that positions certain findings "*dans le faux*" (to distort a Foucauldian idea "*dans le vrai*"). Hard to articulate are injustices that generate credibility for research that unsettles fixed assumptions of plausibility to "science has proved" assertions without asking "whose science," "where and when," and "how"? Equally pressing is the matter of discerning how certain "epistemic identities" and social-structural positionings enable and/or thwart projects of bringing matters of responsible epistemic conduct—of *epistemic responsibility*—into conversation with hermeneutic and testimonial injustice. Many such large, urgent matters have been—and continue to be—the substance of feminist inquiry, whose implications, albeit slowly, unsettle fundamental *ontological* assumptions that shape ways of being and of knowing "who *we* are."

A larger purpose of this inquiry, then, is to consider whether, elaborated and revised, this conceptual apparatus could stretch to inform thinking about the epistemological—and ontological—implications of *climate change skepticism*, where critique has to start from and work against a tenaciously stubborn epistemic-scientific imaginary that is reprehensible in its articulation and harmful in its effects. This critical inquiry would move away from the "singularly" ethical, toward collective, social-political analyses and contestations, where again, the "social imagination" is too slender to inform a sufficiently radical analysis, while Castoriadis's conception of a society that is incongruous with itself opens a way into developing an apt framing discourse for discussion and critique. Furthermore, this issue is less about an imagination that could be individuated than it is about a collective framing—a sedimented yet active *imaginary*—committed to remaining untouchable by accusations of ill-doing and by such ontological questions as "who do we think we are?" These are the most pressing questions.

Implicit in these thoughts is a proposal that relating narratives, counter narratives, can be effective in exposing incongruities that contribute to destabilizing an established social-epistemic imaginary. (Voltaire's *Candide* is, again, a telling example.) Yet complexities persist: the proposal is not that a story, whether factual or fictional, should be taken at face value as a source of knowledge; nor is it reasonable to read stories, even first person narratives, as unequivocally "true." There are many ways to read them: literally, "offering just the facts"; cautiously, recognizing that no single narrative can be expected to do that; phenomenologically, for their part in stripping away, "bracketing" complacent assumptions about how well "we" know/understand/imagine circumstances so radically unlike our own that they only with difficulty find a place—an uneasy place—within the imaginary of having, plenty, availability: heuristically initiating "doxastic shock." Knowing from stories may call for approaching them "blank," with a suspension of belief and disbelief, for even people approached openly and respectfully as informants can be unreliable narrators, can be wary, and ready rather than to manipulate "the truth" than to tell it straight, can simply fail to understand. Nor can engaging epistemologically with particulars—even aberrant particulars—regularly destabilize complacent assumptions in the same way: yet from working by analogy, there is much to be learned. Participants may come away with ignorance/knowledge, in Sullivan's provocative sense: although as she herself shows, ignorance/knowledge, too, can open starting places for working to dislodge an entrenched social-epistemic imaginary. Pedagogically, educationally, engaging with particularity has immense value, for careful teachers and mentors can work with it to generate a just measure of strategic skepticism, which students—and all of us—require in an age of mass manipulation by what poses as "information." Madeleine Bunting's observation about how difficult it is for the "haves" to imagine how it is *not* to have reminds her readers that it is vital to responsible educational practice, whether institutional or "everyday," to educate our imaginations and the imaginations of our students and *other* Others to stretch beyond the familiar: to acknowledge and respect strangeness, to recognize the caution required in treading respectfully on unimaginable territory, to resist assuming to know too much from too narrowly imagining.

That said, this chapter ends without concluding: with ambiguities, puzzles, aporias, for none of these matters comes down to simple either/or choices: thinking about them poses more questions than answers. For a start, it is not always or even usually certain that "the Other," "the oppressed" *wants* to be known "objectively" in her, his or their particularity: people are private, closed and with reasons that demand respect, even despite the implausibility of the "buffered self." This warning is a reminder of the need, in engaging relationally with particularity, to honor "our" opacity to one another, recognizing that even opacity within exposure cannot be conveyed in the formulaic knowledge claims of orthodox Anglo-American epistemology. Honoring opacity requires listening and negotiating across differences, even if these negotiations generate deeper commitments to preserving silence: "to holding open a space for not knowing." Testimony thinly or thickly articulated (in one-liners or extended, elaborated tellings) can be so specific to a situation, to a "who" (singular or plural) in that situation, that if it is to contribute without doing epistemic violence to the intelligibility of the experience told, it needs—perhaps impossibly—to be engaged, and accorded an initial presumption of credibility in its uniqueness, even as hearers and speakers strive to achieve a meeting place in what may be uncharted epistemic territory. Problems are exacerbated by a realization integral to feminist, phenomenological, and other modalities of postmodern thought, that particulars, be they material or human, are neither unmediated in their particularity, nor unmixed in affirming unified transparent wholeness. Nor are they stable, fixed once-and-for-all. So knowing particulars is at once urgent and, in any definitive sense, impossible. Knowledge is commonly unfinished and incomplete, owing to the open-endedness of experiences and meanings, and their time-bound specificities: if our scholarly and pedagogical practice is to be responsible epistemically, we putative knowers need most of all to be cognizant of and open about the limits of our knowledge. We perform an injustice to ourselves and to our readers, students, and colleagues if we resist declaring those limits.

Chapter 5

How to Think Globally, Revisited

Or, A Plea for Ignorance

> Here come I, my name is Jowett
> All there is to know I know it.
> I am Master of this College
> What I know not is not knowledge![1]

In this concluding chapter, I ask how an elusive yet coercive public-philosophical insistence on achieved *certainty* as a *sine qua non* legitimating marker of valid action-generating and policy-informing knowledge, especially but by no means exclusively in ecological-environmental epistemology, ethics, and politics, can truncate scientific and everyday knowledge-seeking in itself and in its public reception.[2] So definitive a veneration of certainty blocks recognition of the ubiquity, the power, and the practical-political effects of acknowledging the "situatedness" of knowledges, again in Donna Haraway's sense. It circumscribes or even disqualifies the most cautiously developed, vigilantly enacted investigative inquiries on whose outcomes ongoing ecological policy and ecologically

---

1. Quip from Benjamin Jowett, master of Balliol College, Oxford University, 1817–1893.
2. This section of the chapter draws on my essay "The Tyranny of Certainty" in *Symposium: Canadian Journal of Continental Philosophy* 21, no. 1 (Spring 2017), pp. 206–218.

informed action rely. The silent impact of such skepticism-promoting venerations of certainty is particularly apparent in US 2012 Republican presidential candidate Mitt Romney's insistent appeal to ongoing scientific *uncertainty* by way of bolstering his adamant refusal to countenance the urgency of "climate change" as a phenomenon whose implications need to inform public environmental/ecological policy and action, and private everyday lives. Similar refusals prevail in numerous affluent societies in the twenty-first-century public world: thus for example, Romney's position reinforces Joni Seager's cautionary observations about "the miasma of uncertainty" tacitly sustained by opponents to environmental-ecological concerns, especially if not exclusively as these concerns affect women's lives and the lives of other Others from an entrenched white affluent masculine norm. The issues are epistemological, ethical-political, and tacitly ontological, with condemnable inaction "justified" in a rhetoric of cautionary appeals to the very uncertainty invoked to block investigative and/or ameliorative actions and policies.

Such cautionary appeals anticipate the participation of "knowing subjects" who are capable—intellectually, situationally, ontologically—of *achieving* certainty and of realizing they have done so. They presuppose a human capacity to recognize and act upon certainty, should putative knowers encounter it. Yet "certainty" is elusive: its achievement may be temporally, situationally, demographically enhanced or enabled to serve extra-epistemological agendas, not all of which are condemnable. Its putative self-evidence as a marker of knowledge achieved—as action-enabling and action-justifying—situates it beyond the critical reach of both scientific and secular inquirers. A "general public," for whom "science" stands as the objective source of certain knowledge is thus vulnerable to assurances that the "evidence," in its *un*certainty, disarms cautionary warnings, renders them irrelevant, excuses inaction. Juxtaposing Haraway's analysis with Donald Brown's ethical-legal analyses of Romney's position is illuminating, for Haraway's insistence that knowledge is always somewhere, and limited, points to a basic epistemic flaw—a cultivated ignorance—on which Romney, if tacitly, relies in his self-excusing rhetoric. In appealing to a quasi-generic lack of certainty in climate science (contrasted with other unspecified ways of knowing), he constructs his position on putatively ethical claims to the effect that it would be unjust to act *without* certainty. Yet Seager convincingly insists: "Scientific uncertainty serves as a refuge for scoundrels of all kinds." Such a "refuge" serves as an escape hatch for failing to act on an epistemically/ecologically responsible "precautionary principle" which invokes obligations to undertake measures designed to prevent environmental harm before a specific danger-threatening event is fully and/or absolutely known—if knowledge so definitive, so certain, is possible. Such is an obligation Brown assigns to Romney, and Seager attributes to threats to human/female health. The ethical-epistemic imperatives are urgent, while certainty remains stubbornly elusive.

In 2011, Mitt Romney opposed US legislation-in-process, intended to reduce greenhouse gas emissions. He cited two principal reasons: first, he did not *know* whether climate change was human caused; and second, since climate change is a *global* problem, he saw no reason for the United States to spend vast sums of time and money to counter it. In consequence, Donald Brown observes, "Mitt Romney's position on human-induced warming is a stunning moral failure." Brown allows that Romney *may not* be claiming there is no evidence of human causation: merely significant *uncertainty* about whether such warming is attributable to human activities. There is a difference: but is it morally relevant, plausible, epistemically responsible? Already in 2007, Brown had maintained: "[I]t is more than 90% certain that observable warming is primarily caused by increasing concentrations of greenhouse gases produced by human activities including the burning of fossil fuels and deforestation" (IPCC, 2007), a view supported by "the most prestigious scientific organizations in the world." This fact, in Brown's view, is morally and epistemically compelling. More than a decade later, the situation is still more urgent. Fundamentally, it poses the *ontological* question that needs urgently, and constantly, to be posed: "*Who* do we think we are?" The answer, I suggest, hinges on an intransigent sense of entitlement in affluent, mainly white, primarily masculine segments of Western-Northern societies: it exposes *the tyranny of certainty* in all its arrogance. To anticipate: an inflated yet unreasoned appeal to "risk" (tacitly, risk to these very populations) is permitted to override implementing a *precautionary* approach where wisdom, reason, and appropriately local—"situated"—knowledges can be enlisted to enhance understandings of specific harms and vulnerabilities, together with cautious, commonable if often tentative, extrapolations to their wider implications.

If blame is to be apportioned, albeit variously, beyond the political climate of the USA and other, putatively white, Western-Northern affluent countries; and if credibility can be claimed for affirming that epistemological theories in professional philosophy can trickle down to shape a going public rhetoric while destabilizing an entrenched social-political imaginary (as I suggest they can), then even though such public debates may not be philosophical in themselves, the enduring aura/influence of logical positivism and its promises is palpable in this veneration of certainty that infuses Romney's rhetoric. Positivism's appeal in the wake of the cultural excesses of the late-nineteenth-century Western world, and of the impact of World War I in the chaos of the early twentieth-century Western world is not difficult to understand; but its erstwhile promise has outrun its inspiration. It has evolved into a form of social-political-epistemic tyranny. *Resistance* to cautious, often epistemologically wise enactments of a precautionary principle in its diverse modalities further contributes to the pull of an illusory *certainty*, contrasted with down-on-the-ground, situated caution in response to overinflated conceptions of "certain knowledge" such as silently (if unwittingly) inform views like Romney's. They find ready uptake in the "prepared mind" of a populace eager to believe certainty is within their/our grasp: a certainty whose attainment is commonly conceived as tantamount to a moral achievement.[3]

With the development of social epistemology in the late twentieth century, in tandem with feminist, racially cognizant, and other postcolonial theories of knowledge, space opens for deliberative-dialogic approaches to knowledge matters that had claimed minimal credence in the positivist era. Here epistemic projects where the situated-ness of knowledges is a given find their place, even as epistemic individualism ceases to be a standard-setting presupposition. Seager's observation that scientific *uncertainty* serves as a refuge for scoundrels is emblematic for thinking about these concerns, especially as it pertains to Romney's position as Donald Brown reads it. Thus, Seager continues, "Chemical-producing and pollution-causing industries have relied for years on the 'cover' that scientific uncertainty affords them." Again, naming Rachel Carson's work as emblematic in the development of respect for precautionary approaches, Seager reads her as offering a way out of what she aptly calls "the closed loop of scientific uncertainty."[4] In fact, this reading of Carson, together with Seager's own inquiries, is germane to understanding the

---

3. The reference to a "prepared mind" is from Louis Pasteur, http://www.access excellence.org/AE/AEC/CC/chance.php. Accessed 6.17.2014.
4. Joni Seager, "Rachel Carson Died of Breast Cancer: The Coming of Age of Feminist Environmentalism," *Signs: Journal of Women in Culture and Society* 28, no. 3 (2003), p. 964.

positive effects of thinking within a precautionary framework, developed and revised in local yet communal deliberations and open, in principle, to contestation and revision. Much ecological thinking then and now occurs in spaces that Carson, if silently, inadvertently, made available. It is she who, in her work and life, conceived and acted upon an unarticulated precautionary principle whose public political and rhetorical effects are still with us in the early decades of the twenty-first century. Perhaps now they are more orthodox in their scientific articulations, but overall, they are no more sophisticated than in her initial enactment. In consequence, even though it has come to be a constant refrain in my thinking, I propose that the urgent question, now, is *whose* certainty or uncertainty is at issue in these confrontations and contestations, and why? Much depends upon the "answers," as epistemic and moral-political issues intersect. They cannot reasonably be held apart.

Central to current Western-Northern debates surrounding such matters is a dangerous stand-off between arguments about risk, and commitments to moving toward enacting precautionary policies and agendas. The latter, perhaps ironically, are widely perceived as being *too cautious*, if by no fixed standards of caution. They too call for case-by-case deliberation. Pertinent in this regard, then, is an observation from Seager's essay "Death by Disease," where she writes: "(The) 'masters of the universe' stance rests on a larger deceit that is . . . deeply infused with gendered social meaning and consequence. . . . Notions of the acceptability of risk are always refracted through a prism of privilege, power, and geography. . . . For whom is 2 degrees warming not dangerous? . . . Who determines what risk is 'acceptable'?"[5] Who lives the consequences of conjectural risks?

Continuous with questions posed throughout this book, this chapter looks more closely at some limitations of "our" knowledge across distances and differences, to argue again for a modest skepticism: an open skepticism, oxymoronic though such a concept may seem to be. The guiding thought is that a healthily skeptical stance, infused with a just measure of *humility*, could curb temptations to pronounce and act in situations where putative "knowers" are insufficiently informed to make responsible, "situated" contributions, despite their being sufficiently knowledgeable for

---

5. Joni Seager, "Death by Degrees: Taking a Feminist Hard Look at the 2 (degrees) Climate Policy," *Kvinder, Køn & Forskning* 3–4 (2009), p. 18.

their views to invite serious consideration. These are delicate issues: they reaffirm the value of epistemic humility, both individual and collective: of admitting ignorance, whether collective or "individual." But none of these are all-or-nothing pronouncements where knowledge and ignorance are starkly opposed and mutually exclusive: most are temporally and situationally bounded. Precautionary recommendations are thus meant to block hasty generalizations based on too-limited evidence, to reserve judgement, to curtail excessive exploratory activity in inadequately known/ understood situations that call for careful thought and wide-ranging consultation: to respect hermeneutic openness. The IDRC situation I discuss in chapter 4 is a pertinent example.

In a 1998 *Hypatia* article, I consider how the then-popular slogan "think globally, act locally" is complicated by the impossibility of affirming universal—or even simply widespread—human circumstantial sameness.[6] In 2019, the contestability of the terms "local" and "global" is still more apparent than in its first articulations. The earlier essay bespeaks a cautious optimism about the slogan's aspirational-inspirational potential, as it infused the "think global, act local" theme of a landmark 1996 International Interdisciplinary Conference on Women, in Adelaide. That theme gestured toward animating transformative, emancipatory epistemic projects within the range of responsible local knowledges (with *their* negotiable limits), thus requiring Western-Northern practitioners to reach imaginatively beyond local boundaries into situations where their everyday beliefs and expectations could be unsettled. The slogan stood as a quasi-mantra, inviting engagement—as it would later be reframed—with "the apparent dilemma posed by the normative demands of cultural relativism when they clash with *our* firmly held moral belief of what is just and right" (my emphasis): calling for a new "theoretical model" to counter a culturally imperialist threat implicit in "the global optimizing strategy of the utilitarian approach . . . reinforced by deontic rationalism."[7] Striking are the false promises that seemed to attend easy recitals of the "act local, think global" mantra, which has to assume a

---

6. Lorraine Code, "How to Think Globally: Stretching the Limits of Imagination," *Hypatia: A Journal of Feminist Philosophy* 13/2 (1998), pp. 73–85. Reprint in Uma Narayan and Sandra Harding, eds., *Decentering the Center: Philosophy for a Multicultural, Postcolonial, and Feminist World* (Bloomington: Indiana University Press, 2000).

7. I quote the letter of invitation sent to participants in the symposium "Cultural Relativism and Global Feminism,"1997 APA Pacific Division conference, which built further on this theme.

certain incontestability of both *the local* and *the global* if it is to assume that *the global* will fit within the imaginative capacity of locally aware white affluent Western-Northern theorists and activists.

While local (Western) activism may, on occasion, succeed in enhancing and/or refining practice-informing knowledge, while it may strengthen and consolidate white Western conceptions of justice and right, the risk of venturing too far is constant. It cautions against entering regions—cultural, geographical, discursive, personal—where theorists and activists know not whereof they/we speak. Vigilance against epistemic/hermeneutic imperialism is an ever-urgent imperative, as attempts to enact local principles globally confront the limits of their salience in situations beyond the reach of locally knowledgeable imaginations, be they collective or "individual." The risk of enacting epistemic violence is constant, especially where cultural relativism is invoked to affirm local immunity, as in (largely Western) warnings to "back off" from condemning practices that appear to enforce female degradation and exploitation in cultures at home and abroad that differ radically from "our own," and where "our own" allegedly stands as the unmarked, uncontested, normative indicator that everything is in order as it stands. Female genital mutilation is an obvious—if overinvoked—example of entrenched local practices that violate (white Western, and multiple other) feminist principles.

Hermeneutic Humility

In 2008, in "Advocacy, Negotiation, and the Politics of Unknowing,"[8] I ask how far locally achieved medical/health care knowledge and practice can travel across cultural-social distances while maintaining aptness, credibility and efficacy. The exploration was prompted by the 1990s Canadian IDRC[9] project in Tanzania I discuss in chapter 4, whose purpose was to "consider certain epistemic demands integral to knowing well across differences from tacitly assumed or explicitly instantiated social norms: differences of gender, race, class, place, circumstance, demography, culture, inextricably intertwined and reciprocally constitutive." It asks how a locally entrenched social-epistemic imaginary can be hospitable

---

8. Lorraine Code, "Advocacy, Negotiation, and the Politics of Unknowing." *Southern Journal of Philosophy* 46, no 1 (2008), pp. 32–51.

9. IDRC refers to Canada's International Development Research Centre.

to innovative endeavors from elsewhere: endeavors that aim to enable knowing well across social-political-cultural differences: *radical* differences. Why would an entrenched imaginary erect obstacles? As I have noted, the IDRC's aim was to turn "an entire health system around, moving it from ossified methods of gathering, evaluating, and circulating knowledge and tired old administrative practices and distributions of epistemic power and authority, toward a responsive, responsible, democratic complex of social-natural epistemic interactions."[10] Pursuing that aim calls for collaborative-interpretive epistemic approaches, attentive to empirical factuality in its multiple modalities, and to specificities—local, affective, demographic, gendered, which are neither clear nor self-evident to social-cultural "outsiders." Although well-crafted examples can contribute to seeing such projects through, although they can generate debate from which new understandings may derive, in such situations, empirical findings may not—perhaps cannot—speak for themselves. They require engaged communication, interpretation; they need time.

Knowing/understanding across radical differences, therefore, requires more than a capacity to move around in the language of "the other." As I suggest in chapter 2, it calls for an approach akin to Maria Lugones's "world-travelling,"[11] which is far removed from "eleven-countries-in-eleven days" tourism. This is a traveling prepared to spend the time and to practice the openness and humility integral to eschewing sedimented perceptual/interpretive habits and prejudices; to remove (again following Ludwig Wittgenstein) the "pair of glasses on our nose through which we see whatever we look at. It [normally] never occurs to us to take them off."[12] *Humility*, cast as an epistemic virtue yet not in a groveling Uriah Heap sense, is vital to performing this function well—as it is for participants in the IDRC project. It marks a courageous engagement: respectful, receptive, standing back to deliberate, to consult, to acknowledge errors, to engage in further research; to think again. Having discerned such virtues in Carson's epistemic practices—in her ways of understanding "nature"—Joni Seager comments: "'Humility' often has religious overtones, but Carson . . . is speaking to the rash overconfidence

---

10. Code, "Advocacy," p. 35.

11. Maria Lugones, "Playfulness, 'World' Travelling, and Loving Perception," *Hypatia* 2, no. 2 (1987), pp. 3–19.

12. Ludwig Wittgenstein, *Philosophical Investigations*, trans. G. E. M. Anscombe (Oxford: Basil Blackwell, 1968), #103.

of humans who act as though they can remove themselves from the inescapable truth that humans are part of nature."[13] Such overconfidence encourages superimposing ideologically infused readings of social-political situations and diagnostic practices onto events in ways that are opaque to practitioners in radically different circumstances: readings that fail to fit within the purview of their taken-for-granted everyday expectations.

The point is not to condemn advocacy or ideology *tout court*, but to insist that ideologies, too, must come before "the bar of reason." Hence, in the Tanzanian example discussed in chapter 4, the central claim is not that symptoms of a disease must differ radically across geographical-cultural-social circumstances, but that *living* them can be culturally incommensurate, diversely challenging, across situational/ontological differences. Recognizing such lack of fit could be impossible from superficial observations, cursory pronouncements, diagnoses that invoke putatively univocal conceptions of "abnormality" in attempting to squeeze diverse events and populations into standard boxes: from arrogant perception. Humility calls for a readiness to enter situations where it could be a mistake—a failure of responsible practice—to suppose that discrete circumstances will be instantaneously self-explanatory, *knowable*, owing to the putatively generic expertise of the investigators. It manifests in the conduct of inquirers who are prepared to stand back; to "reserve judgement"; or again, in Karen Barad's words, to "meet the universe half way." Doing so requires multifaceted analyses: interpretive, hermeneutical, empirical, where these putative "strands" are not distinct and separate one from the others but interactive: *intra-active*, in Barad's rich sense.[14] Good interpretive practices take time: a requirement that is as urgently *epistemic* as it is budgetary. Even in relatively affluent twenty-first-century circumstances, it can be difficult if not impossible to fulfil, leaving epistemologists, diagnosticians and lay people helplessly bewildered. Yet postpositivist theorists and practitioners in the twenty-first century may have no choice but to work in situations of "manufactured uncertainty," neither of *their* making nor within their capacity to undo. Nor can time reasonably be allocated to processes of investigation from one-size-fits-all before-the-fact presuppositions: from convictions that received epistemic values and practices will be effective across situational/populational/structural/conceptual differences, with no

---

13. Seager, "Rachel Carson Died of Breast Cancer," p. 965.
14. Karen Barad, *Meeting the Universe Halfway: Quantum Physics and the Entanglement of Matter and Meaning* (Durham, NC: Duke University Press, 2007).

good reasons to affirm their specificity. While these "situational" factors may seem to be "beside the point" in philosophical theories of knowledge, ignoring them widens a gap between theory and practice whose consequences can be epistemically damaging: can truncate the potential of socially responsive/responsible approaches to knowledge. Thus in *Ecological Thinking*, I commend Karen Messing's diagnostic practices in her studies of workplace health, where she shows how insufficiently *specific* diagnostic criteria: criteria that draw upon large, general epistemological theories and practices can fail to uncover particular—perhaps even unique—symptoms that seem to "break the rules." Such events call for time-consuming interventions, even as too narrow investigative samples and too hasty conclusions likewise limit the scope and results of empirical inquiry.[15] Messing studies how "work stations, tools, and equipment designed for 'the average man' tend to place many women and some men at a disadvantage: their effects in pain, stress and discomfort are rarely discernible except in the language of failing to meet an abstract, one-size-fits-all criterion of fitness."[16] Appropriate treatment requires taking particular yet contrary-to-the-norm events seriously: refraining from dismissing them as transient anomalies; being attentive to their temporal/spatial/demographic specificities. Time and money feature prominently among the intransigent obstacles to knowing such situations well enough: informed negotiation is essential to these debates.

In the IDRC's Tanzania project discussed in chapter four, the investigators' departure from a top-down epistemology of mastery, enacted in moving toward responsive and responsible democratic practices of social-natural epistemic interaction, is especially noteworthy. These practices were marked by a gradually learned sensitivity to habitus and *ethos*, and by redistributions of epistemic authority across populations and places. Hitherto "appropriate" distributions of funds and services had drawn on "one size fits all" assumptions: an approach that involved dispatching equal sums of money and identical packages of drugs to each district. Often, the fit was not good enough because the *knowing* was not good enough. It required time, patience, and collaborative deliberation for investigators to discern why measures effective elsewhere were

---

15. Karen Messing, *One-eyed Science: Occupational Health and Women Workers* (Philadelphia: Temple University Press, 1998).

16. I quote my paraphrase of Messing in *Ecological Thinking*, p. 54.

less so, here; required the investigators to discard fixed assumptions of ontological-situational homogeneity across Tanzanian populations: a point more generally pertinent across twenty-first-century social-epistemological inquiry. Hermeneutic hesitation—integral to humility—had consistently to be practiced if investigators were to uncover and respond fittingly to particularities occluded in the too-hasty generalizations informing policy and practice. Rarely could this investigative practice remain "individual," solitary: consultation and discussion were vital to its going through. Here, engaging actively with locally endorsed practices was imperative for enabling investigators to learn to see past the sedimented habits of inquiry blocking the way to sensitive responses: this was the project's purpose. Nonetheless, where time and money are urgent concerns, the power of the most carefully crafted examples developed to illustrate and elaborate complex epistemological matters may be not be realizable, for lack of adequate resources. Such a thought is epistemologically damaging, despite its resolution often being beyond the resources of investigators, and to the detriment of knowledge-seeking.

As I also note in chapter 4, sketching the outlines of this story neither affirms the "power of example" nor shows that appeals to example consistently generate worthy epistemic outcomes. Sometimes they do. Equally, such readings can fail to produce mind-changing effects, despite the capacity of apt examples to point toward viable, innovative ways of "going on." *Certainty* about an example's effectiveness is not always— indeed, is rarely—within reach. Nonetheless, catalyzed by appropriate examples, hermeneutic engagement can enhance the powers of epistemic challenge to interrupt and inform thought and action. Still, cautionary comments are in order: incorporating *analogical thinking*/reasoning into this mix offers one potentially fertile, if complex possibility. It is a delicate but invaluable practice: drawing a pertinent analogy—"it's just like falling off a log"—can offer encouragement or generate misinterpretation—especially with an analogy more specific than the minimal one just cited. Too often, simple empirical examples and counter-examples fall short of making their point clear. Analogical thinking—"well, it's rather like exploring a new city"—can move understanding forward. Like examples *tout court*, analogies are crafted to function as points of entry into understanding events, biases, situations. In complex, multi-faceted situations, their signature strength is in their hermeneutic openness: in the place they create for cooperative deliberation and humility as contributors to understanding, negotiating, rethinking: in resisting premature closure.

The IDRC's mandate was to investigate causes of deaths in specific areas of rural Tanzania: to know "thickly," not in formulaic ("check the relevant boxes") detail, *how* deaths from malaria occurred across diverse parts of the country and its populations; how they were understood and addressed. Citing my Spindel (2007) paper: "one of the most intransigent obstacles in the power/knowledge complex these reversals and consequent 'successes' have encountered has been an entrenched reliance on stereotypes embedded in the instituted social imaginary . . . and notoriously resistant to counterevidence . . . through which administrators and other outsiders purport to know local populations."[17] This claim connects with thoughts about a lack of humility, about too-readily imported and assigned labels, about interpretations that expose failures to pause and assess the aptness of fit: practices essential to 'learning from example.' Epistemologically, such practices can be difficult to enact in cultures where speed and time are of the essence, blurring the hermeneutic power of exchanges that require more extended hearings if they are to enhance understanding across differences and sameness. Troubling likewise are practices of reading simplified empirical/observational claims as paradigmatic examples for "knowledge in general," whereas such claims more than likely represent a *kind* of knowing whose reach is truncated in consequence of the minimal ways of "going on" they inevitably enable.

"Knowing populations"—knowing people and events "well enough" within population-specific, situational-cultural locations—cannot be a one-off, "individual" happening: it takes time, deliberation, critique. Yet thoughtfully developed and enacted knowing practices can learn from judicious—appropriately humble—appeals to example. The contrast is with generalizing from too few factual or fictional particulars: making the strange familiar; reading newly encountered people and situations as fitting too easily into pre-formed categories, ready-at-hand stereotypes—of women, men, children, white or nonwhite people, the old, the foreign, the "lesser" class or race—and truncating inquiry and understanding in so doing. ("Isn't that just like a woman?!") Knowing from well-narrated examples and counter-examples carries a potential to shake the rigidity of stereotype-sustained typologies and categories, to draw people into "what if?" situations, with potentially commendable

---

17. Code, "Advocacy," p. 40.

consequences, usually with caution. Yet invoking examples to dislodge stereotypes is emphatically not tantamount to generalizing from too few particulars: well-elaborated, fitting examples are particularly effective in countering the putative "necessity" of patterns of behavior in policy decisions: instructive admonitions to those seeking understanding. At its best, it is a cautious, caring process in the attention it gives to the specificities of its tellings, and in its dialogic openness to ongoing deliberation. Its potential to foster affective engagement is one of its most notable aspects.

Kristie Dotson, responding to my Spindel paper in the *Southern Journal of Philosophy* (discussed in chapter 4), invokes powerful evidence contesting the strength and interpretive capacities of narrated examples and counter-examples such as those I recount. She deplores projects purporting to advance understanding across differences that too swiftly superimpose examples onto events where the fit is less than clear. Her analysis could catalyze a critical-constructive debate that would draw readers into what Harvey Cormier, citing William James, calls "Ever not quite," "[which] has to be said of the attempts made anywhere in the universe at attaining all-inclusiveness."[18] These thoughts notwithstanding, Dotson's "In Search of Tanzania" article is rightly critical of the position I present: I have learned from it. She shows how epistemic friction can—and must—spark radical rethinking and action, without denying that it can also disrupt, destroy such responses.

Inevitable incongruities notwithstanding, effective practices of invoking narrated examples and counterexamples come with a potential to expose something of the philosophical and everyday complexity of interpretive projects across notable differences: to highlight their successes and failures, to open ways to respectful contestation in understanding complex situations through specific yet open-ended interpretive-hermeneutic frameworks.[19] Well-crafted examples point toward diverse ways of understanding and/or interrogating events, situations, human interactions: of moving toward evaluating engaged practices repeatable, *mutatis mutandi*, in situations sufficiently analogous to find epistemic

---

18. Harvey Cormier, "Ever Not Quite: Unfinished Theories, Unfinished Societies, and Pragmatism," in Sullivan and Tuana, eds., *Race and Epistemologies of Ignorance*, pp. 59–76.

19. Germane to these thoughts is Heidi Grasswick and Nancy A. McHugh, eds., *The Power of Example* (Albany: SUNY Press, in press).

resources in the knowings they detail. They caution against a hubris enacted in too-hasty assumptions of sameness and difference: again, they call for humility, and they take time. Still more significantly, in well-functioning exchanges apt examples/counter-examples can work effectively to encourage wariness of "observer" analyses that stand too far apart from their subject matters and fail, in consequence, to listen carefully, participate critically or constructively, probe deeply enough to engage well with contradictions and uncertainties: to practice humility.

Pressing now, in the early decades of the twenty-first century, is an urgency these debates generate about knowing *whose* "firmly held moral belief(s)" are at stake in discourses about justice and rights: about how to engage such beliefs; to analyse the sources of their warrant and trace their genealogies. These are difficult issues, especially in the wake of a (Western-Northern Anglophone) epistemological history of projects whose goal is to determine *formal* "necessary and sufficient conditions" for the existence of knowledge "in general," uncluttered by the messiness of specificity and detail. Setting aside any thoughts about cultural relativism, if "global optimizing" is read as a project of achieving maximum (economic-social) homogeneity and uniformly "efficient" material productivity across human and "natural" resources, then it has been relentlessly complicit in promoting cultural imperialism: in expanding the limits of a hegemonic instituted epistemic imaginary of the self-certainties of a white Western-Northern affluent androcentric world. Admittedly, this rereading of global optimizing is impressionistic and programmatic, loosely structured and inconclusive. It tries to say everything at once, not to construct a linear argument; it challenges the very possibility of its own project—albeit from within a diffuse feminist- and postcolonial-infused approach. It derives in large part from a conviction that it is virtually impossible to think *globally* in any but a vaguely gestural sense, even in a media-saturated world where with one click, everything can be right there in front of "us." It may also, depending on how theorists pin it down, *localize* it, or *locate* it, be impossible even to think *locally*, responsibly and well, for "we" (whose extension is contentious) too often fail to perceive/see/realize the local either as "merely" local or as carrying larger global import. In a different register, "we" often fail to take adequately into account the conditions of that "world's" locality, of what has made it possible, and how: to engage critically with *its* genealogy and/or the social-epistemic imaginary that holds it in place.

Strategic Ignorance

In the year 2000, my 1998 *Hypatia* essay was reprinted in the volume *Decentering the Center*,[20] which was, then, an apt title. Now, more than two decades later it generates questions analogous to those posed by the "think global, act local" slogan. In consequence of radical shifts in the politics of knowledge in the first decades of the twenty-first century, with their ongoing exposure of global and selectively local ignorance; and in view of innovatively unsettling—if mainly positive—developments in feminist and antiracist theory and practice in these same decades, the very idea of "the center" is increasingly elusive, and shifting. A new starting point is clearly required. Such a beginning could be akin to the aftermath of a quasi-Cartesian radical doubting; a phenomenological bracketing; or what Charles Mills aptly calls performing "an operation of Brechtian defamiliarization, estrangement, on [y]our cognition."[21] Mills's recommendation comes from his suspicions of "ideal ethical theory" and the dislocated presuppositions on which it rests. And, while this thought will not be new to feminist or to other nontraditional epistemologists, taking it *seriously enough* involves recognizing it as a significant component of responsible epistemic agency, now, across a range of issues where the explicit aim is to *know*, responsibly and in its existential-ecological specificities, the extent of "our" ignorance. Such "estrangement"—such acknowledgment of ignorance—need not paralyse inquiry. Recall a challenge early naysayers posed to Genevieve Lloyd's *The Man of Reason* asking, now that she had deconstructed the pretensions of Reason, what was she proposing to put in its place?[22] Her apt response was to the effect that, given how long it had taken to understand the variable, changing historical intermappings of reason and masculinity, it would be facile, premature, and indeed irresponsible simply to offer up a new construct in their place. Yet equipped with the questionings and understandings Lloyd's analyses have made available, and mindful of their

---

20. Uma Narayan and Sandra Harding, eds., *Decentering the Center: Philosophy for a Multicultural, Postcolonial, and Feminist World* (Bloomington: Indiana University Press, 2000.)

21. Charles Mills, "'Ideal Theory' as Ideology," *Hypatia: A Journal of Feminist Philosophy* 20, no. 3 (2005), p. 169.

22. Genevieve Lloyd, *The Man of Reason: 'Male' and 'Female' in Western Philosophy*, 2nd ed. (London: Routledge, 1994).

implications, feminist and other postcolonial philosophical inquiry at their best have proceeded with new, provocatively cautionary assessments of their own local character.

An "estrangement," a bracketing project, seems thus to be of a piece with what certain essays in the 2007 *Hypatia* "Ignorance" volume, in Sullivan's and Tuana's *Race and Epistemologies of Ignorance*, and in the *Agnotology* volume, have taught us.[23] It amounts, provocatively, even naughtily, to a plea for ignorance: paradoxically, to an insistence on needing to *know our ignorance* in order to engage well with the *aporias* of our times. Responding adequately to such a plea does not require disingenuous disavowals analogous to those white Western women were, historically, well trained to utter in epistemic deference to the superior cognitive powers of the men of their time and station. Yet it points toward ways of counteracting the arrogance of white Western perceptions (thinking of Marilyn Frye[24]) while proceeding, if the lesson is well learned, with a renewed but not a deferential humility. It is in part about acknowledging and countering white ignorance. But here, following Alison Bailey, it is not just about knowing and deploring injustices done, but about learning—in her words—from "strategic uses of ignorance by people of color," achievable, she maintains, not by moving out from the local with its presuppositions and its logic intact, but "by learning to think in new logics . . . developing (following Maria Lugones) an account of subjectivity that centers on multiplicity,"[25], thereby turning sharply away from the abstract individualism of classical liberal ethics and epistemology.

---

23. Now-classic texts are *Hypatia: A Journal of Feminist Philosophy*, Special Issue: *Feminist Epistemologies of Ignorance* 21, no. 3 (2006); Shannon Sullivan and Nancy Tuana, eds., *Race and Epistemologies of Ignorance* (Albany: SUNY Press, 2007); Robert Proctor and Londa Schiebinger, eds., *Agnotology: The Making & Unmaking of Ignorance* (Stanford, CA: Stanford University Press, 2008). See also Shannon Sullivan, *Revealing Whiteness: The Unconscious Habits of Racial Privilege* (Bloomington: Indiana University Press, 2006) and *Good White People: The Problem with Middle-Class White Anti-Racism* (Albany: SUNY Press, 2014).

24. Marilyn Frye, "In and Out of Harm's Way: Arrogance and Love," in her *The Politics of Reality: Essays in Feminist Theory* (Trumansburg, NY: The Crossing Press, 1983), pp. 52–83.

25. Alison Bailey, "Strategic Ignorance," In Sullivan and Tuana, eds., *Race and Epistemologies of Ignorance*, pp. 81–82; Alison Bailey, "White Talk as a Barrier to Understanding the Problem of Whiteness," in George Yancy, ed., *White Self-Criticality Beyond Anti-Racism: What Is It Like to Be a White Problem?* (Lanham, Maryland: Lexington Books, 2015), pp. 37–57.

Strategic ignorance (elaborated in ways Bailey might now resist) can function as a heuristic for this new beginning: can work to confront, to engage with multiple and diverse structures of what Shannon Sullivan has called ignorance/knowledge—an idea which, for her, "does not collapse ignorance and knowledge into one another . . . [but] denies, or at least places under suspicion, the purported self-mastery and self-transparency of knowledge, as if nothing properly escaped its grasp. . . . It helps one to peek behind knowledge . . . to see what un-knowledges help to compose it and upon which [it] depends." With Sullivan, I am suggesting that among the urgent issues facing all of "us" now, if variously in our diverse situations, is a need to engage in projects committed to uncovering ignorance-passing-for-knowledge—where ignorance can be understood as "an active production of particular kinds of knowledges for various social and political purposes"[26] and where "uncovering" will be a brave, even a dangerous process.

Ignorance itself is no unified, univocal conceptual framing or practice: in its most neutral aspects, it amounts simply to not knowing. In less benign manifestations it can be cultivated, coerced to serve putatively nefarious ends—such as keeping populations, singly or collectively, under-informed, hence incapable of thinking or deciding wisely. Ignorance connects with the "manufactured uncertainty" that is the title of this book, which is intended to point toward its manipulative effects. There is something oxymoronic in suggesting that ignorance could be a consequence of certain kinds of knowledge production, but examples abound. To select just one, current debates about climate change, informed by allegedly "well warranted evidence," too often cloak the financial interests that generate approval for one solution over others. Saboteurs need to be knowledgeable if they are to sabotage successfully. This "something" makes the endeavor still more challenging. Now-classic examples abound. Among the most striking are what Robert Proctor names "the tobacco industry's efforts to manufacture doubts about the hazards of smoking"; Oreskes and Conway's analysis of how, "in the hands of the Marshall Institute and those it has influenced, climate science has been profoundly misrepresented and a great deal of confusion and ignorance produced"; and David Michaels's tracking how "the pharmaceutical industry is devoting sizable resources to conducting studies whose results will increase sales, but will not necessarily provide the information physicians need to select

---

26. This and the previous quote are from Sullivan, "White Ignorance," p. 154.

the best drug for their patients."[27] A full list of examples of ignorance thus generated would be very long. It would be provisional, time-bound, and incomplete, but its purposes are clear: manufacturing uncertainty ranks high among the commitments and projects of such knowledge producing and ignorance tracking practices.

Within the dominant social-epistemic imaginary of the twenty-first century white Western capitalist world, I have noted, certain taken-for-granted assumptions nurture an all-too-human tendency to confer credibility and epistemic authority almost exclusively upon knowledge/ignorance produced and circulated by accredited public-institutional bodies and/or eminently credentialed thinkers and scholars. None of these thoughts are new, but the pervasiveness of the practices, and the public-corporate obstacles held in place to block the potential credibility of counter-narratives and of "differently credentialed" knowers show that coming to know our ignorance is no small project. Hence, with reference to scientific communities, Bruno Latour speaks in favor of narrative analyses, claiming that "the only way to respect . . . heterogeneity and . . . locality is . . . to do *a lot* of philosophy. But philosophy is not unifying factors . . . [it] is a *protection* against the hegemony of the present sciences."[28] His observation pertains well beyond "the present sciences" to encompass protections against the hegemony of the ignorance/knowledge that informs rigid conceptions of the "local" and the "global" and more, in the twenty-first century.

In this regard, there is epistemic traction to be achieved from engaging with "bio-regional narratives," a practice I invoke in *Ecological Thinking*, with thanks to Jim Cheney. Narratives "grounded in geography rather than in a linear, essentialized narrative self" are vital, Cheney argues, to developing an ethics—and an epistemology—of accountability.[29] They map local relations to discern conditions for living mutually sustaining lives within a specific locality—be it an institution of knowledge production, an urban or a rural setting, a workplace, a geographical region,

---

27. Proctor and Schiebinger, eds., *Agnotology*, p. 11; Oreskes and Conway, *Merchants of Doubt*, p. 81; Michaels, *Doubt is Their Product*, p. 98.

28. Bruno Latour, "Irréduction," in Werner Callebaut, *Taking the Naturalistic Turn, or How Real Philosophy of Science Is Done* (Chicago: University of Chicago Press, 1993), p. 218.

29. Jim Cheney, "Postmodern Environmental Ethics: Ethics as Bioregional Narrative," *Environmental Ethics* 11, no. 2 (1989), pp. 117–34.

a wilderness, a body of water, a community, culture, society, state, and/ or the interrelations among them. Here I am conceiving "geography" richly, broadening its everyday extension to map demographics, populations, social practices, and power relations ecologically, a conception that draws, if tacitly, on affinities with human—not physical—geographers. The strength of such narratives is in the detail of their situational sensitivity, their genealogical (power-infused) exposure of local thriving or deteriorating, of supportive or oppressive conditions, much as some of the essays in the *Agnotology* volume expose the strategies and conflicted interests that collaborate in "the making and unmaking of ignorance" in the bio-regions they examine. The weakness of such narratives is in the risk of their achieving merely local pertinence: of imagining a bio-region as closed, internally harmonized, static, and representative, hence as an implausible surrogate for any other region. Yet narratives, too, require critical evaluation in the specificities of their internal detail and their external interconnections with wider, diverse power-saturated systems. Such sober cautionary reminders need not detract from their epistemic potential.

Clearly, as Latour insists, analyses of this kind must proceed cautiously, with no unrealistic hopes of *purity* (contra Lugones), and conjecturally, toward developing and circulating hypotheses, protests, proposals for discussion, deliberation, refutation. They construct complex multi-dimensional demographic mappings of place or locality, where the forces, lines of influence and interest, trajectories, intersections, frictions that infuse and shape inquiry are mapped narratively, to arrive at textured understandings of resources, influences, conflicts and harmonies, interdependencies, ignorance and knowledge. The outcome may well be a layered, multilayered interpenetrating tale—clearly not a tale told by an idiot, nor by only one teller, nor even in unison by many tellers—but by many voices in diverse times and places, affirming, contradicting, contesting, reinforcing one another. The result is a complex image/metaphor, neither legible nor audible all at once, nor by one abstract "individual," not part of any monologic, punctiform 'S knows that p' pretension to know, but a place where ecologically informed, deliberative-dialogic attention to multiple particularities, congruencies, frictions has much to offer.

The challenge in mapping and evaluating such projects is to craft a conceptual apparatus sufficiently sensitive and powerful to carry the inquiry forward. Here, recall Gilles Deleuze's and Felix Guattari's deceptively simple definition of philosophy as "the art of forming, inventing, and fabricating concepts"; "the discipline that involves *creating* concepts."[30] With this definition, rather than returning philosophy to the aridity of mid-twentieth-century Anglophone conceptual analysis, Deleuze and Guattari remind their readers that effectively transformative philosophical projects put critical-constructive conceptual apparatuses into circulation: apparatuses sufficiently rich and complex to generate innovative ways of inhabiting, engaging, understanding the physical and the human-animate world. It need scarcely be added that feminist, ecological, postcolonial, materialist, and critical race theory together with so-called "new" social theories are just such sustained, overlapping projects of critiquing and displacing conceptual structures that have long held oppressive social orders in place, uncontested. Epistemologically elaborated, bio-regional narratives, too, map knowledge-enhancing and knowledge-impeding structures and forces, structures of ignorance and knowing, with the aim of deriving normative conclusions that—deliberatively, negotiably—could translate from region to region, perhaps not without remainder, but as instructively in the disanalogies they expose as in the analogies they propose. Here, an innovative conceptual repertoire is required: literary works afford one viable resource.

Just such an example of an ordinary ignorance failing to see itself for what it is occurs in the portrayal of a woman called Maureen, the hitherto affluent white South African protagonist of Nadine Gordimer's 1981 novel *July's People*. She, in her everyday life, takes universal human sameness unquestioningly for granted: sameness of relationships and feelings, of conjugal arrangements and gendered divisions of labor, of the significance of places and objects. Yet despite her avowed commitment to acquiring a sense of how it is for her black African servant, July, and for the people of his village where he provides refuge for her and for her family from racial riots in the city—when she is uprooted from that life to take refuge in the village—she persists in these entrenched assumptions. For her, Gordimer writes, "The human creed depended on validities staked on a belief in the absolute nature of intimate relationships between human beings. If people don't all experience emotional satisfaction and deprivation in the same way, what claim can there be

---

30. Gilles Deleuze and Felix Guattari, *What Is Philosophy?* trans. Graham Burchell and Hugh Tomlinson (London: Verso, 1994), pp. 2, 5. I draw attention to these thoughts in *Ecological Thinking*, pp. 26, 27.

for equality of need?"³¹ she wonders. Even when she is relocated at a distance from the taken-for-granted certainties of her then life, she cannot recognize the specificity of her conceptions of sexual loyalty, "suburban adultery" and love to the white middle-class Anglophone society where she has learned them: she cannot move, self-critically, to wonder whether these putatively universal human verities might not count as universal, after all. So apparently small a conceptual move is beyond the scope of her imagining. In offering a reading of this novel my aim is, in part, to show how little this white woman is able to/allows herself to realize of the sheer *local* character of the local, even in human intimacy: how ill-placed and indeed ill-advised she is to make of that "local" a touchstone from which to imagine how it is to know the world from his standpoint, for July, her erstwhile black servant, her "boy." (Bailey notes, "Ignorance flourishes when we confine our movements, thoughts, and actions to those worlds, social circles, and logics where we are most comfortable."³²) A bracketing, a quasi-Cartesian move such as I propose might have served this woman well: had she learned to realize how narrow the range of the local really is/was, she might have been better able to see the presumptuousness of merely enlarging its scope and terms of reference to explain the less local, the hitherto more remote, now right before her eyes. She fails to understand the value of engaging bio-regionally with July and with "his place": of constructing a narrative that would require her/enable her to know something of how it is for him and his people. That failure is ultimately her undoing.

In 1997, I read "Women Workers and Capitalist Scripts," by Chandra Mohanty³³ for its capacity to stretch its readers' imaginations *not* by homogenizing or aggregating the strange and the familiar, but by drawing circumscribed quasi-global conclusions from sensitive local mappings, in critical, cross-cultural comparisons. There, Mohanty proposes

---

31. Nadine Gordimer, *July's People* (New York and London: Penguin Books, 1981), pp. 64, 103. I discuss this novel in greater detail in "'They Treated Him Well': Fact, Fiction, and the Politics of Knowledge," in Heidi Grasswick, ed., *Feminist Epistemology and Philosophy of Science: Power in Knowledge* (Dordrecht: Springer, 2011), pp. 205–22.

32. Bailey, "Strategic Ignorance," p. 91.

33. Chandra Talpade Mohanty, "Women Workers and Capitalist Scripts: Ideologies of Domination, Common Interests, and the Politics of Solidarity," in *Feminist Genealogies, Colonial Legacies, Democratic Futures*, ed. M. Jacqui Alexander and Chandra Talpade Mohanty (New York: Routledge, 1997), pp. 3–29.

that understanding local specificities requires "pay[ing] attention to the commonalities of their/our common *and* different histories," opposing "ahistorical notions of the common experience, exploitation, or strength of Third-World women or between third- and first-world women, which . . . naturalize normative Western feminist categories of self and other."[34] Affinities with the Tanzanian example I invoke in chapter 4 will again be apparent here. Anticipating Bailey's "Strategic Ignorance" arguments and her " 'White Talk' As a Barrier to Understanding the Problem with Whiteness" from a different point of critical entry, Mohanty proposes that "the predicament of poor working women and their experiences of survival and resistance in the creation of new organizational forms to earn a living and improve their daily lives . . . offers new possibilities for struggle and action."[35] Her mappings point toward a "geographically anchored" methodology capable of showing how "the local and the global are indeed connected through parallel, contradictory, and sometimes converging relations . . . which position women in different and similar locations as workers."[36] Such intersecting lines of critical evaluation abandon both epistemic deductivism and "ideal epistemic theory" to opt for inductive, interpretive readings, sensitive to particularities: readings that eschew reductive approaches, to advocate working phenomenologically with/from a plurality of analyses and methods. A reader might object that quasi-narrative examples such as those cited here are merely that—stories, tales, narratives: they can claim no epistemological pertinence. Such an objection is neither fanciful nor foolish; it may even be apt. At times story-telling could slide into the anecdotal and lose epistemic force in so doing. But my thought is that—especially for social epistemologists and those committed to hermeneutic, interpretive inquiry—definitive epistemic one-liners rarely enhance the knowledge that contributes to understanding: to knowing "our" way about the world. Narratives can open ways into thinking further about these matters, to contesting reasonably, to assenting with understanding.

What recommends Mohanty's agenda is the manner of inquiry she employs, the positioning of the inquirer's voice, its practice of letting the evidence speak—neither naively (as if without contest) nor by slotting it into ready-made boxes. She enjoins responsible listening, suggests

---

34. Mohanty, "Women Workers," p. 28.
35. Mohanty, "Women Workers," p. 23.
36. Mohanty, "Women Workers," p. 6.

connections across social-cultural positioning and stops short of others, asks for vigilance against pushing the limits too far, even of imaginative knowing. These may sound merely like trite, common-sense admonitions; but such vigilant listening has not characterized the utilitarian strategies of global optimizing, which manifest scant concern for ecological effects, whether local or global, literal or metaphorical. It keeps space open for arguing forcibly against cultural imperialism.

My plea for ignorance, running variously through these conjectures, is not about claiming Charles Mills is wrong to deplore white ignorance or Shannon Sullivan to ask, with evident dismay, *why* she knows so little about Puerto Rico. It amounts, variously, to a plea for vigilance to recognize, address, and endeavor to counter strategies that are tacitly or overtly committed to the making and sustaining of ignorance. It recommends a certain skepticism—a strategic, Pyrrhonian skepticism—in the face of increasing social-political awareness of how little we, whoever we are, really know about most of what we need to know in the social-political world, in its multiple instantiations. Ignorance is often but not always regrettable, yet acknowledging our ignorance, the limitations of our knowledge, can be an epistemically responsible action, be it individual or collective. Ignorance unrecognized, unacknowledged, unknown is not merely deplorable, but dangerous. Thus, returning to an earlier point in this chapter, projects inspired by the slogan "think globally, act locally" but conducted in ignorance are potentially dangerous in their power to enact damaging epistemic injustice across events, protagonists, situations. Yet my intention is less to conclude these thoughts with a counsel of despair than to advocate a counsel of circumspection at a historical moment when it is no exaggeration to suggest that all of "our" knowledge is exposed in its instability, in the narrowness of its scope, the arrogance of its purview, the unknowings that hold it in place. For reasons such as these, in his provocatively titled *Empires of Belief: Why We Need More Scepticism and Doubt in the Twenty-first Century*, Stuart Sim recommends skepticism as "a cure for philosophical pretension, a permanent internal critique." He proposes: "It is in the conflict with the world's grand narratives that skepticism performs its greatest service . . . raising doubts about the grounds on which claims are made, systems organised, authority assumed, and power wielded. It becomes a critique, in other words, of the drive toward certainty."[37] Only if such a

---

37. Stuart Sim, *Empires of Belief: Why We Need More Scepticism and Doubt in the Twenty-first Century* (Edinburgh: Edinburgh University Press, 2006), p. 41.

drive can be monitored, set aside, collectively not just individually, only if projects of inquiry can be conducted to the accompaniment of such ongoing "internal critique," will it be responsible to attempt to "think globally, act locally."

Intellectual Virtue

My purpose, now, is to reaffirm that, in my view, the pivotal, overriding intellectual imperative and virtue is *epistemic responsibility*. This claim is both large and minimal: large in suggesting that epistemic responsibility might encompass all epistemic virtues, minimal in proposing that cognitive-intellectual virtues could, possibly, reduce to one and that "one" would establish itself by overtly or covertly eschewing affect as a vital contributor to knowledge sought and achieved. There seem to be no universal, formal criteria for judging an act of knowing "epistemically responsible" and few exemplary cases: multivocal deliberative and cooperative discussions/judgments provide the best, perhaps the only, plausible resources in such cases. Nonetheless, from granting that ethical and epistemological matters are reciprocally constitutive, it seems to follow that epistemic responsibility, collectively enacted, must occupy a central place in virtue epistemology as practice and theory, and that virtue epistemology is all of these—an ontological, ethical, political, affect-infused epistemic position and practice. The minimal guiding thought, then, is that knowing well is a fundamental social, individual, political obligation for people who would live well with/for themselves and with others, in most imaginable circumstances, be those circumstances individual, collective, or an amalgam thereof, be they benign, malign, or neutral. With Anglo-American epistemology's "empirical simples," fulfilling such obligations seems to be so matter-of-course, so trivial, as to require little argument to invoke attentiveness and/or to require caring, especially in materially replete societies where anyone—leaving the extension of the term vague—can know about cups on tables or cats on mats. But there are circumstances where even such knowledge cannot and indeed *should not* be taken for granted: with just slightly more complex empirical examples, assumptions about the ubiquitous accessibility of the stuff of which knowledge is made will be more presumptuous than realistic.

Convictions about the uniform, ubiquitous accessibility of the stuff of which knowledge is made are seriously contested in feminist, postcolonial, antiracist inquiries, together with a taken-for-granted interchangeability of epistemic and moral-political subjects and of the "stuff" to be known. Such "interchangeability" cannot be presumed innocent, or even possible, before the fact of its enactments. Once emphasis shifts from requiring a view from nowhere to knowing the social-political-affective implications of knowing "facts" *somewhere*, in epistemically diverse situations, it is incongruous to presume without contest that autonomous (white) man, alike to all other putative knowers in all cognitive circumstances, with equivalent access to the objects and practices to be known, stands as the default moral-epistemic subject; or that knowledge is an individual achievement or possession. Nor, in consequence of current thinking about reciprocal intra-actions (following Karen Barad) between thinking-knowing and feeling, can the stark aridity of traditional postpositivist empiricism claim epistemic power and unqualified respect.

In the years since I brought the concept into Anglophone philosophical discourse in my eponymous 1987 book, *epistemic responsibility* has gone through multiple iterations, contestations, and variations not just in my work, but elsewhere, especially—if tacitly—in social epistemology and in hermeneutically oriented engagements with knowledge questions.[38] Despite the idea's having been something of a "sleeper" in mainstream Anglophone epistemology since the time of the book's publication, the virtue of responsible and responsive epistemic conduct retains a vital salience. It is currently claiming a renewed salience, for virtue epistemology and, if tacitly, with local caveats and variations, for epistemology more widely conceived. In the wake of early twenty-first-century developments in Anglo-American social epistemology, where its responsive aspect figures centrally in the relationality often definitive of the quality of epistemic interactions, that salience is still greater than it was in 1987.

Thinking about virtue epistemology historically, as a precursor to twenty-first-century social epistemology, recall that for Aristotle, virtue *simpliciter* is "such a . . . state as makes a man good and able to perform his proper function well."[39] It predisposes its practitioners toward reasonably dependable realizations of certain valued ends: collectively,

---

38. See my "Epistemic Responsibility," in Ian James Kidd, José Medina, and Gaile Pohlhaus, Jr eds., *The Routledge Handbook of Epistemic Injustice* (London: Routledge 2017), pp. 89–99.

39. Aristotle, *Nichomachean Ethics*, trans. J.E.C. Weldon (London: Macmillan, 1927), Book II, ch. 4.

in open societies, it is a *sine qua non* precondition for a viable social order. Responsible epistemic conduct, both collective and individual, is thus central to virtuous epistemic lives, and epistemic activity (i.e., *knowing*) is itself constitutive of viable ways of being in the world. Hence, assessments of human character (singular, dyadic, collective) and of the epistemic climate of a family, society, community, institution of learning must consider the quality, not just the quantity, of such activity, in its diverse local contingencies. Virtuous epistemic conduct is manifested less in *how much* would-be knowers know, more in *how well* they know: how well they, collectively, enact such knowing; how carefully they honor and respond to uncertainty. In short, a knower's and/or a community's intellectual goodness is not just about having consistently good scores in knowledge-seeking/-constructing projects that "come out right"; it is about cultivating dependable, worthy epistemic habits, capacities, and qualities in larger virtue-orientated relations to the social, political, physical, and material world. Virtue has constantly to be cultivated: it manifests in commitments to the social, collective practices that define or contest it, circular though this thought is. It manifests in the doing, not just in the analyses. Individually, it is an ongoing, renewing project in a Sartrean existential sense, not a fixed attribute that triggers automatic responses.

Collectively—in "collectives" as small as intimate relationships, or as large as a city, a community, a political movement, or a society—its nature and boundaries are works in progress. Intelligence, then, is manifested in commitments to engage and to know/understand situations clearly, carefully, and with just measures of wisdom and affect. In the twenty-first-century "first world," intelligent knowers are rarely judged virtuous for being unmoved or unaffected by what they seek to know: apt feelings, appropriate responses are integral to virtuous, wise epistemic conduct, both individual and collective.[40] Prudence is a different matter: it is vital yet troubling. Thinking of the risks involved, the persistence required in endeavoring to know well recalls Michel Foucault's *aude sapere!*: "dare to know," or "have the courage, the *audacity*, to know."[41]

---

40. I allude to the title, if not the substance, of Allan Gibbard, *Wise Choices, Apt Feelings: A Theory of Normative Judgment* (Cambridge, MA: Harvard University Press, 1990).

41. Michel Foucault, "What Is Enlightenment?," trans. Catherine Porter, in Paul Rabinow, ed., *The Foucault Reader* (New York: Pantheon Books, 1984), p. 35.

Prudence would thus involve cultivating a just sense of how much/how far it is appropriate or wise to dare; excessive prudence could manifest in epistemic timidity, inertia, excessively conservative epistemic lives concerned more with avoiding error or with not looking foolish than with the creativity required to explore untrodden paths, experiment with bold ideas, support contentious yet worthy causes. It could prevail in cautious, not-innovative, education practices and systems. Like the Socratic gadfly, courageous knowers—singly or collectively—keep an epistemic community on its toes: they exhibit virtue in so doing.

In a non-technical sense, intellectual goodness manifests in "realist" orientations, exhibited in doing justice to the object(s) of knowledge, be they animate or inanimate, physical, personal, collective, material, theoretical, ideational. It stretches from knowing people to knowing theories, places, histories, populations, literary works, toxic or nontoxic substances: to any potential objects of knowledge. It requires humility and understanding yet it does not fear or resist truth, even as it eschews epistemic arrogance. Rarely is the aim to fulfill these commitments in solitary isolation: commonly, with subject matters that matter, coming to know responsibly is the (perhaps interim) product of ongoing educational practices and communal, social living. Virtuous knowing manifests in readiness to grant a fair hearing, to develop and sustain good judgments, but not in commitments to tolerate any and every form of trivia, malice, or nonsense. It is difficult to defend this last claim without reading it as a plea for an austere earnestness and against playfulness, comedy, fabrication, or folklore: difficult also to acquire and maintain a sensitivity to differences among prudence, consistency, integrity on one hand, and dogmatic intransigence on the other. Something akin to an Aristotelian mean becomes a guiding principle after all, and attributions of epistemic virtue will be made communally, often deliberatively, and situationally in a sense related to, if not precisely equivalent to Donna Haraway's "situated knowledges."[42]

These are some of the requirements of virtuous knowing. Putative knowers cannot be presumed before the fact to be situated alike, or even comparably, in the social-political-geographical places where knowledge

---

42. Donna Haraway, "'Situated Knowledges,' The Science Question in Feminism and the Privilege of Partial Perspective," in her *Simians, Cyborgs, and Women: The Reinvention of Nature* (NY: Routledge, 1991), pp. 183–202.

is made and circulated, and where its effects are unevenly distributed in and for human lives. As Susan Babbitt's analysis suggests, questions about the politics of knowledge—questions about intellectual virtue and the risks of doing epistemic injustice to people, places and practices—cannot adequately be addressed through a formal, abstract, impersonal rubric. They require rich yet careful phenomenological-affective engagement with the particularities of people, places, and experiences, seasoned with a just estimation of the scope and limits of "our" understanding, and with an openness to encountering the unfamiliar, the strange while resisting temptations to "fit it into" preset taxonomies and frames of reference. Such are among the requirements of virtuous knowing. Putative knowers cannot be presumed before the fact to be situated alike, or even comparably, in the social-political-geographical places where knowledge is made and circulated and where its effects are unevenly distributed in and for human lives. Emulating the humility Carson practises, Haraway's "modest witness" would take cognizance of such considerations and in so doing (justly) gain a reputation for the wisdom integral to epistemic responsibility.

While this sketch may appear to endorse a picture of intellectual virtue as an austere, even prim, individual achievement, such is not its intention. A move from "individual" to commonality, to community, with respect to virtue may seem incongruous since virtues have tended to be conceived as practices of "individuals": my virtues cannot be yours, nor yours mine. And thinking should move in that direction: from individual to social, with "the social" conceived as an amalgam of discrete "individuals." But there are reasons for thinking otherwise. If the ethical and epistemic *sensibility* manifested in responsible conduct is conceived as inculcated in practices of social nurturing and training, moving from the communal to the individual is not difficult to conceive: it is more plausible in "real life" situations than in moving from individual to community. People develop ethical sensibilities from being embedded within historically and culturally specific, familial and wider social, diversely populated ways of life. Such a position is eloquently articulated by Annette Baier, in the mid-1980s: "A person, perhaps, is best seen as one who was long enough dependent upon other persons to acquire the essential arts of personhood. Persons essentially are *second* persons."[43] (Miranda Fricker reads *epistemic*

---

43. Annette Baier, "Cartesian Persons" in her *Postures of the Mind: Essays on Mind and Morals* (Minneapolis: University of Minnesota Press, 1985), pp. 74–92, at p. 82.

sensibility as a product of socialization "a social training of the interpretative and affective attitudes in play when we are told things by other people."⁴⁴) For educators, parents, activists, to mention the most obvious few, assumptions prevail that learning can instill such virtues, or trade upon their infectious properties, albeit not in the direct way one would administer a medication or share a bar of chocolate.

Significantly, therefore, it would be implausible to propose that Jean-Paul Sartre's "self-taught man" exemplifies intellectual virtue/epistemic responsibility in consequence of his voraciously, inexhaustibly collecting items of information, related to one another only by their listing in alphabetical order in a single reference text. A misbegotten conviction informs his putative project to the effect that it is possible to know everything while understanding nothing. But attributions of intellectual virtue are rarely if ever plausible on the basis of single, isolated acts of knowing well: an inveterate prevaricator cannot wipe the slate clean with one impressive act of truth-telling. At least, some evidence of wider understanding, some estimation of the relative significance and implications of an action, are integral to a virtue-centered epistemology, and are components of intellectually virtuous practice. Juli Eflin captures the point: affirming the teleological character of the virtue-centered epistemology she advocates, for which *understanding* is an overall goal, she writes: "[I]t is not individual, unrelated facts that I want to pile up, especially not trivial facts—even if they do meet the necessary and sufficient conditions for knowledge. . . . I want important, interrelated facts . . . skills that enable me to learn more . . . a coherent framework into which new information can fit and cohere. . . . I want understanding and the ability to increase my understanding in the areas I deem important."⁴⁵ Yet there is a caveat: although the central figure

---

Citing Baier, I elaborate this idea in chapter 3, "Second Persons" of *What Can She Know? Feminist Theory and the Construction of Knowledge* (Ithaca: Cornell University Press, 1991). (Curiously, Stephen Darwall, in an exhaustive list of citations in *The Second-Person Standpoint* [Harvard, 2006], makes reference neither to Baier's nor to my part in introducing and developing this idea.)

44. Fricker, "Epistemic Injustice." p. 145.

45. Juli Eflin, "Epistemic Presuppositions and Their Consequences," in Brady and Pritchard, eds., *Moral and Epistemic Virtues* (Oxford: Wiley-Blackwell, 2004), p. 49. Here I draw also on my "Virtue, Reason, and Wisdom," in Stan Van Hooft, ed., *Handbook of Virtue Ethics* (New York: Routledge, 2013), pp. 188–99.

in such an epistemology is "the inquirer," as perhaps it must be, and an inquirer who is specifically located and knows "from there," Elfin's inquirer seems to acquire intellectual virtue more individualistically than would the more explicitly socially-communally engaged inquirer on whom this picture of responsible epistemic agency is centered: a thought that limits but does not deny her/his intellectual stature.

In short, neither intellectual virtue nor epistemic responsibility is primarily about individual cognitive activity. These derive from and are attributable to communities of inquiry, human institutions, and communally created-enacted practices. Such enactments can go awry to issue in practices and instill sensibilities that are more vicious than virtuous: violations that occur within a social-epistemic imaginary infused with and constituted from power-infused and biased, ethically-politically reprehensible beliefs and prejudices, also socially instilled. The sedimented racism, sexism, homophobia, and multifaceted ignorance endemic to and upheld within social groups, in certain times and places, are glaring examples. Their stubborn intransigence highlights a further complexity of virtue-based epistemology and moral-political theory: an elusive need for criteria of external evaluation; for judgments of what seems to be enclosed and internally justifying. In an epistemologically well-functioning society, a significant majority of virtue-attesting epistemic interactions may be based in trust. Matter-of-course as such a claim could be for members of benign and reasonably safe, affluent, smoothly functioning social "kinds," societies, or parts thereof, in this possibility resides a potential for epistemic injustice, for violations consequent upon condemnable socially-politically instilled beliefs and abuses of trust. Nor can epistemically reprehensible conduct be exonerated with the simplistic excuse of its being merely a product of putative knowers' socialization. It is not clear that even a situation analogous to Neurath's raft could supply the vantage point from which judgments of the whole could be articulated and/or enacted.

Whose virtue? Whose wisdom? These references to intellectual virtue "in general" appear to take for granted that virtue (as with "the natural light reason" for Descartes[46]) is alike in all men, all virtuous people are men, and all are alike in their virtuous attributes and practices. None of

---

46. Genevieve Lloyd, *The Man of Reason: "Male" and "Female" in Western Philosophy*, 2nd ed. (London: Routledge, 1994). Lloyd writes that for Descartes, "this natural light of Reason is supposedly equal in all" (p. 44).

these assumptions is tenable in the twenty-first century: historically, their putative universality derived from locally entrenched, rarely contested or justified philosophical assumptions of universal human sameness (albeit sameness within an unnamed group). As Genevieve Lloyd amply demonstrates, reason, wisdom, intelligence, prudence are concepts whose essence, derivation, and significance are neither written in stone nor in any other enduring, situationally neutral medium. Yet the purpose of considering the implications of this caveat, whether following Lloyd or for inquiry "in general," is emphatically not to propose that there are certain virtues for men and other, usually lesser, virtues for women; virtues for white folks and lesser virtues for not-white folks, despite such suppositions' being (silently) entrenched in quotidian practices and philosophical presuppositions of societies, classes, ages, and races throughout recorded (western) history. The conceptual content of reason, and likewise of intellectual and moral virtue, as they have evolved through the history of Western philosophy derives, again as Lloyd shows, from locally, temporally contingent ideals of intellectual conduct deemed virtuous for men—where "men" too has a frequently, if imperceptibly, narrow extension.

Minimally, some level of epistemic interdependence is required for participation in a functioning epistemic and broader society or community. If people could never count on one another in knowledge-as-information exchanges, it is unclear how viable societies, institutions, and human relationships could persist.[47] Demographically, perhaps coincidentally, white Western-Northern philosophers seem to have lived and claimed membership in classes, genders, and races where such assumptions and expectations hold, where epistemic interdependence can be regarded as a matter of course; and whose violations are straightforward wrongs. But consciousness-raising practices since the 1960s and 1970s in Western/ Northern societies have moved toward a renewed focus on gender, class, race, ethnicity, together with multiple and intersecting aspects of sameness and diversity in philosophical inquiry and in everyday discourse. They have disrupted tacit convictions about a uniformity that countenances injustices

---

47. In *Epistemic Responsibility*, I cite Peter Winch: "The notion of a society in which there is a language but in which truth-telling is not regarded as the norm is a self-contradictory one." Peter Winch, "Nature and Convention," in *Ethics and Action* (London: Routledge and Kegan Paul, 1972), p. 61. Whether this view remains plausible in the twenty-first century is a different, and vital, question.

of omission and commission in sustaining the benign figure of "the man of reason": the autonomous (tacitly male, tacitly white) epistemic agent.

Unsettling for sedimented assumptions that accord this figure ethical and epistemological pride of place are Susan Babbitt's apt observations, as she asks her readers to wonder not only why philosophy is so consistently masculine, but equally urgently, "why Philosophy is so white."[48] For Babbitt, the question of whiteness connects with going conceptions of "the nature of Philosophy as the pursuit of wisdom." A contrast she draws between constitutive presuppositions of autonomy, self-realization, and freedom in North American societies, where "equality for women is defined in terms of what men have within the current society," and the social equality enacted by Igbo women of east Nigeria,[49] informs her exposure of systemic injustices that Western assumptions enact and perpetuate, in their pretentions to universality. These injustices attest to a perniciously irresponsible social-epistemic infrastructure: an instituted social imaginary[50] manifested in sustained failures to question, and thence perhaps to know, the exclusionary effects of the governing ontological assumptions, not just of Western *philosophy*, but also of its trickle-down effects in people's everyday lives. The presumptuousness of taking for granted the ubiquitous accessibility of the stuff of which knowledge is made gestures toward a simpler version of this thought.

Babbitt aptly contends that "if professional philosophers, because of a commitment to certain views, are unable to raise the sorts of questions that critically identify deep-seated assumptions about who we are

---

48. Susan Babbitt, "Philosophy's Whiteness and the Loss of Wisdom," in George Yancy, ed., *The Center Must Not Hold: White Women Philosophers on the Whiteness of Philosophy* (Lexington Books, 2010), p. 169. For Babbitt, Philosophy (capital "P") refers to philosophy taught in English-speaking universities in the USA, Canada, and the UK. (See fn 1, p. 190.) She does not rest her case on one reading, but even so telling an exception undermines pretensions to universality implicit in hegemonic Western conceptions of reason and wisdom.

49. Babbitt, "Philosophy's Whiteness," p. 172. Babbitt draws extensively and impressively on Nkiru Nzegwu, *Family Matters: Feminist Concepts in African Philosophy of Culture* (Albany: SUNY Press, 2006) for her discussion of Igbo culture.

50. The concept of a social imaginary is from Cornelius Castoriadis. See his *Philosophy, Politics, Autonomy: Essays in Political Philosophy*, ed. David Ames (New York: Oxford University Press, 1991), p. 62; and "Radical Imagination and the Social Instituting Imaginary," in Gillian Robinson and John Rundell, eds. *Rethinking Imagination: Culture and Creativity* (London: Routledge, 1994).

as human beings, we might wonder whether the dominant academic practice of Philosophy, at least in the English-speaking traditions, is really about wisdom after all. It might be . . . about correctness."[51] Her analysis makes clear that the "essential arts of personhood," too, are always specifically located—socially, culturally, racially, and by gender. It resonates with Sandra Bartky's exploration of "psychological oppression," where she addresses the psychic alienation Franz Fanon, living in a white society, experienced as "the estrangement of separating off a person from some of the essential attributes of personhood."[52] (Miranda Fricker reads Fanon, contentiously, as showing how even "functioning as an informant on everyday matters . . . [entails] being accepted as a compatriot in the community of the rational."[53]) It is hard to claim recognition as a practitioner of the virtues, be they ethical, intellectual, or other, when a person's very being falls beneath the radar of social-ontological acknowledgment. For untold numbers of women and other Others in the white Western-Northern world, such acceptance has rarely been a matter of course: my comments about ubiquitous accessibility adumbrate this point. The thought recurs forcefully in Alexis Shotwell's analysis of "appropriate subjects," where (with reference to Charles Mills's *Blackness Visible*) she notes that, for Ralph Ellison's *Invisible Man*, the basic Cartesian problem about the self—about the *sum* of the *cogito*, as she neatly puts it—originates not (as for Descartes) in doubting and then reaffirming his own existence, but in "being socially created as sub-human . . . [being] ontologically subject to the power of a gaze that denies his existence while holding social power."[54] When social-ontological homogeneity (or even contiguity) cannot be taken for granted, it becomes clear that diversity is not just about "kinds," but about multiple positionings in and in relation to structures of power and privilege as well.

---

51. Babbitt, "Philosophy's Whiteness," pp. 169–70.

52. Cited by Miranda Fricker, in her "Epistemic Injustice and a Role for Virtue in the Politics of Knowing," in Brady and Pritchard, eds., *Moral and Epistemic Virtues*, pp. 156–57.

53. Fricker "Epistemic Injustice," p. 157.

54. Alexis Shotwell, "Appropriate Subjects: Whiteness and the Discipline of Philosophy," in Yancy, ed., *The Center Must Not Hold*, p. 123.

Now, *after virtue*[55] in its neo-Aristotelian white Anglo-American Western sense has too-long claimed uncontested, polite hegemony in the very philosophical circles Babbitt refers to, it is time to concur enthusiastically with her proposal that "the pursuit of wisdom requires a kind of humility."[56] Intellectual humility is a central ingredient of responsible epistemic practice, while its opposite—epistemic arrogance—is just such an ingredient in situations of epistemic oppression and injustice. Revisiting Aristotle's definition of virtue, as "such a . . . state as makes a man good and able to perform his proper function well," confirms how vital it is to understand how such functions are historically, culturally, racially, and gender-specifically assigned, defined, inculcated, praised, or condemned, and not innocently, neutrally so.

Social epistemology claims its title in large measure from the centrality it accords to testimony and to knowledge-conveying exchanges between and among people in the real world.[57] Striking is the attention many social epistemologists, now, accord to situated examples of epistemic negotiation and deliberation; likewise for the subtle, far-reaching effects of a linguistic shift from impersonal, third-person propositional claims that "S knows that p," to the language of speakers and hearers. It is still more striking for its late appearance in an epistemological landscape where a presumption in favor of analyzing formal propositional examples has created and maintained a remarkable distance between people's epistemic lives and the kinds of example philosophy can legitimately address: a distance that contributes to the minimal attention epistemologists before social epistemology were conceptually equipped to accord to questions about responsible, innovative—virtuous—epistemic conduct. Thus, where reliance on the *S-knows-that-p* rubric enabled formal Anglo-American theories of the twentieth century to transcend the vicissitudes of the world in specifying a priori, necessary and sufficient conditions for "knowledge in general," social epistemologists in their multiplicity return to and reclaim the world, both human and other-than-human, with its incoherence and messiness, its contradictions and specificities, to engage with real-world,

---

55. The reference is to Alasdair MacIntyre, *After Virtue: A Study in Moral Theory* (Notre Dame, IN: University of Notre Dame Press, 1984).

56. Babbitt, "Philosophy's Whiteness," p. 170.

57. I elaborate these thoughts in "Testimony, Advocacy, Ignorance: Thinking Ecologically about Social Knowledge," in Alan Millar, Adrian Haddock and Duncan Pritchard, eds., *Social Epistemology* (Oxford: Oxford University Press, 2010), pp. 29–50.

quotidian epistemic interactions and negotiations. In consequence, the very idea of "knowledge in general" is drained of content.

Many of the issues social epistemology generates have contributed to blurring the lines that had long separated epistemology from ethical and political debate and influence: lines drawn to ensure that politics, ethics, and other human interests could not block the route to epistemic objectivity. Descriptively and normatively, epistemology was to be a disinterested pursuit. Now, for many social epistemologists, ethical-political questions—about trust, power, advocacy, negotiation, epistemic community, the ethics of belief—inevitably enter the discourse: not, as was feared, to the detriment—but to the enhancement—of responsible, compassionate inquiry. Acts of giving and receiving testimony commonly, if tacitly, involve such issues. They matter in human lives, but their mattering need not/should not obliterate possibilities of engaging responsibly, virtuously, critically, even at times objectively, with them. Showing how such engagements can be achieved opens spaces for thicker, more wide-ranging analyses of responsible epistemic conduct than could be accommodated in the astringent versions of postpositivism. In 2006, and with such thoughts in mind I concluded the introduction to *Ecological Thinking*:

> Ecological thinking is not simply thinking *about* ecology or *about* "the environment": it generates revisioned modes of engagement with knowledge, subjectivity, politics, ethics, science, citizenship and agency, which pervade and reconfigure theory and practice alike. It is committed to "meeting the universe halfway." First and foremost a thoughtful practice, thinking ecologically carries with it a large measure of responsibility—to know somehow more *carefully* than single surface readings allow. It might be difficult to imagine how it could translate into wider issues of citizenship and politics, but the answer, both simple and profound, is that ecological thinking is about imagining, crafting, articulating, endeavouring to enact principles of ideal cohabitation.[58]

This thought still captures my working conception of the project of epistemology.

---

58. Lorraine Code, *Ecological Thinking: The Politics of Epistemic Location* (New York: Oxford University Press, 2006), p. 24.

While I draw on Aristotle for the defining principle of intellectual virtue, here I suggest that as epistemology has evolved since the end of the twentieth century, with testimony currently accorded a central a place among sources of knowledge, so too *epistemic responsibility* has come more clearly into its own: it requires radical rethinking across a range of ideas and issues hitherto judged *hors de question* in thinking about knowledge. In this shift, especially in social epistemology, away from an epistemology of monological pronouncements and punctiform knowledge claims quintessentially exemplified in the "S knows that *p*" formula, and toward a language of speakers and hearers, the question "Whose knowledge is at issue?" acquires a new urgency and pertinence, and for two principal reasons. First, with reference to an issue I discuss in *What Can She Know?*, even in philosophy, persons can no longer plausibly be regarded as discrete, self-contained, isolated "individuals"/entities. From Annette Baier's apt reminder that "persons are essentially second persons," it follows that epistemic responsibility must be an interactive-communicative virtue from the get-go, so to speak, even when it is practiced and enacted solitarily, vis-à-vis singular objects of knowledge, events, states of affairs. So long as knowers are represented merely as interchangeable place-holders whose knowing in ideal observation conditions is the basis of epistemological analysis, the very idea of epistemic responsibility can gain no intellectual traction: once the products of their activity are awarded the honorific status "knowledge," there is no sense of choices being involved in making attributions of responsibility. Second, situations that require epistemically responsible responses can no longer be approached before the fact as replicas or elaborations of one-off face-to-face encounters with an "object" of knowledge. Knowing responsibly becomes a complex, nuanced interactive/intra-active project that calls for responsiveness, humility, and care. These intellectual virtues assume prominence as knowledge production shifts toward becoming a collaborative, negotiated project reliant on the participation of many voices, not all of them speaking in unison.

Babbitt's apt observation that "the pursuit of wisdom requires a kind of humility" suggests a way forward in thinking about responsibility as it can shape epistemic conduct, not only in interpersonal interactions, but in scientific and other modalities of engagement with the larger world. For some thinkers, undoubtedly, the very idea of humility will be repellent for its hitherto obsequious associations with such figures as Uriah Heap in Charles Dickens's *David Copperfield*, whose groveling, hypocritically self-effacing demeanor may have contributed to a common view that humility carries something of an unpleasant odor, is a less than admirable characteristic. I attempt to dispel such an impression by showing humility at work, so to speak, in the research practices of Rachel Carson and, in a different context, of Donna Haraway, before proposing how Babbitt's observation can play an emblematic part in articulating the implications of wise—hence virtuous—critical engagement with philosophy's whiteness.

Epistemic Responsibility and Virtuous Knowing Today

In *Ecological Thinking*, I position Carson as an exemplary figure for her thinking about responsible epistemic practices in scientific inquiry and in her interactions with the social-political environment of her time.[59] She advocates intellectual-moral humility in scientific inquiry and in scientifically informed practices, to displace what she perceives as the hubris that drives indiscriminate pesticide use: the arrogance, in her view, of human aspirations, without humility, to achieve "control of nature." She argues, in effect, for a shift toward taking respectful account of nature's putative integrity and of the creatures whose lives and habitats are irrevocably damaged by such practices. Although Carson's work is directed specifically toward ecological practices in "nature," I suggest intellectual humility, with variations in content according to the subject matters involved, is a widely salient epistemic virtue, especially when it is parsed to take cognizance of its alignment with respectful, careful attention to diversity across even apparently identical situations and populations. Carson's work anticipates Donna Haraway's figure of the modest witness whose engagement with "heterogeneous histories" introduces critical consciousness, committed to unsettling deep-seated social preoccupations with preserving the same.[60] Modesty thus conceived— which I read as contiguous with humility—entails wariness of any rush to judgment, of too-swift attributions of homogeneity across situations, populations, feelings and attitudes, thoughtless applications of ready-made

---

59. *Ecological Thinking*, pp. 39–40. I draw on this passage in the following observations.
60. Donna J. Haraway, *Modest_Witness@Second_Millennium.Female_Man_Meets_ OncoMouse: Feminism and Technoscience* (New York: Routledge, 1997), see e.g. p. 51.

taxonomies. In its virtuous modalities it entails a respectful, yet neither cold nor cloying, attentiveness to similarities and differences.

Not unlike Aristotelian virtues, humility and modesty are to be practiced according to a mean; nor are they merely "individually" pertinent. Excessive modesty and excessive humility recall the cringing posture of Uriah Heap: epistemically they appear to tell against responding to Foucault's *aude sapere!* challenge. So, Julia Driver rightly notes "a moral virtue like modesty may involve epistemic vice, since the modest person underestimates self-worth (to some small degree), and is thus making a mistake."[61] This apt caution does not gainsay the significance of modesty or humility as germane to responsible inquiry; rather it enhances their constitutive part in thoughtful practices that comprise virtue-cognizant approaches, broadly conceived, while avoiding excessive piety.

As Babbitt's reflections suggest, questions about the politics of knowledge—questions about intellectual virtue and about the risks of doing epistemic injustice to people, places and practices—cannot adequately be addressed through a formal, abstract, impersonal rubric. They require rich yet careful phenomenological-affective engagements with particularities of people, places, and experiences, seasoned with a just estimation of the scope and limits of "our" understanding, and with an openness to encountering the unfamiliar, the strange while resisting temptations to "fit it" into preset taxonomies and frames of reference. Such are among the requirements of virtuous knowing. Putative knowers cannot be presumed before the fact to be situated alike, or even comparably, in the social-political-geographical places where knowledge is made and circulated and where its effects are unevenly distributed in and for human lives. Emulating the humility Carson practises, Haraway's "modest witness" would take cognizance of such considerations and in so doing (justly) enhance her/his reputation for the wisdom integral to responsible epistemic conduct.

Pertinent, then, is Nelson's provocative question: "Is Dismissing the Precautionary Principle the Manly Thing to Do?"[62] Deploring widespread resistance to taking precautionary measures seriously, Nelson notes that

---

61. Julia Driver, "The Conflation of Moral and Epistemic Virtue," in Brady and Pritchard, eds., *Moral and Epistemic Virtues*, p. 106.

62. Julie A. Nelson, "Is Dismissing the Precautionary Principle the Manly Thing to Do? Gender and the Economics of Environmental Protection," *Ethics and the Environment* 20, no. 1 (Mar. 2015), pp. 99–122.

"although some economists are joining the calls for dramatic action on climate change . . . the leading voices in U.S. policy—and because of U.S. power, important voices in international negotiations—speak from a position that assumes that we are basically in control of our situation, and have no need for attitudes of care or caution."[63] The precautionary principle denotes a duty to prevent harm when it is within *our* power to do so, *even when no evidence is in*. This principle has been codified in several international treaties to which Canada is a signatory. Domestic law makes reference to this principle, but implementation remains limited. The "tyranny of certainty," to which I have referred, assists nay-sayers in their denial projects as they over-emphasize what has not been achieved, thereby minimizing what indeed has been gained/brought to visibility and awareness, and has opened the way for ongoing inquiry and interpretation.

On the way to a conclusion, these final observations are guided by the apt title, and the text itself, of Stephen L. Esquith's 2010 book, *The Political Responsibilities of Everyday Bystanders*, and by reading it together with Claire Hemmings's 2010 article "Affective Solidarity: Feminist Reflexivity and Political Solidarity"[64] to which I refer in chapter 1. Because both of these works point clearly toward thinking about the power of examples, and about their capacity to initiate, contribute to, challenge, or discredit settled modes of social-political thinking, now, I cite their titles in full. Esquith's concern is with bringing everyday bystanders—folk, us—to "recognize and meet their shared and institutional political responsibilities for severe violence."[65] My purpose in drawing on his text has been to show—by example—how would-be knowers might read examples, variously, for their epistemic potential as contributors to interpretive, hermeneutic understandings and thoughtful actions across a range of knowings, in late twentieth-/early twenty-first-century century inquiry. My interest is analogous, if less specifically centered: it derives from diverse attempts to articulate, to understand such *epistemic responsibilities* as pervade everyday human lives in the twenty-first-century Western/Northern world, to initiate new or renewed realizations of responsibilities often unacknowledged, yet imperceptibly powerful. Thus,

---

63. Nelson, "Is Dismissing . . . ?" p. 101.
64. Claire Hemmings, "Affective Solidarity: Feminist Reflexivity and Political Solidarity," *Feminist Theory* 13 no. 2 (2012), pp. 147–61.
65. Stephen L. Esquith, *The Political Responsibilities of Everyday Bystanders* (University Park: Pennsylvania State University Press 2010), p. 1.

when soon-to-graduate philosophy students in an urban Canadian university know virtually nothing of climate change and global warming, of the social-political implications of World War II, of thalidomide, of the racism, sexism, and xenophobia that pervade their society, one must ask how they will engage, as worthy participants and/or as "everyday bystanders," with responsibilities in/to the societies where they live or to other societies where equivalently urgent issues arise, societies made by these and other events. As points of entry into these discussions, examples—well-narrated and deliberated examples—again, not just one-liners—contribute, gradually, to closing such gaps.

"Bystanders" need also to learn how to read, listen, deliberate: to look critically, hermeneutically—beyond the "given," where examples can be read as self-contained within the time and place of their making: thus irrelevant to "us." Pedagogically, it is a constant challenge to determine how to "start from where students are"—and from where a wide range of innumerable other interlocutors are: the power of intelligently crafted examples, held open to deliberation, to discussion, to challenge and counterchallenge offers invaluable resources. Hemmings contributes a salient piece to such thoughts, arguing for the epistemic significance of the affective dissonance that feminist politics *"necessarily begins from,"*[66] and of the power that infuses/immobilizes such dissonance. The *frictions* that dissonances generate can unsettle taken-for-granted presuppositions and cognitive habits. Well-articulated, such frictions can initiate rethinking, reconsidering. Equally, they can prompt retreat to a stubbornly intransigent status quo. These are among the ever-present tensions in innovative epistemic projects. Thus Hemmings's commitment to drawing on the "affective" joins with certain endeavors in Western/Northern feminist philosophy to restore the legitimacy of care and concern to epistemic practices, moving toward reclaiming respectability for discussions of why and how people care/should care about knowing well, again recalling Vrinda Dalmiya's apt question "Why should a knower care?"[67]

If this question pertains only to such contrived exercises as follow upon asking "What is it like to be a bat?" there might appear to be no reason to care. But engaging with complex modalities of thinking, feeling, being, together with according epistemic value to *empathy* as is

---

66. Hemmings, "Affective solidarity," p. 148.
67. Vrinda Dalmiya, "Why Should a Knower Care?" *Hypatia: A Journal of Feminist Philosophy* 17, no. 1 (2002), pp. 34–52.

common, if diversely, to Esquith's and Hemmings's thinking, offers a route toward responses richer than the narrow repertoire on offer from simpler exercises of affirmation or denial. Analogous claims pertain to such works as Beauvoir's *She Came to Stay*, Camus's *The Stranger*, Sartre's *Nausea*, and their twenty-first-century counterparts: novels that draw readers into thinking phenomenologically, empathically, experientially, *with* other Others, and within their explicitly situated circumstances. Such claims contribute to understanding not just intellectually, but affectively and in (potential) solidarity. They are integral to the philosophical projects of their time and still inform the innovative implications of ours.

Bibliography

Adams, Douglas. *Life, the Universe and Everything*. Pan Books, 1986.
Agamben, Giorgio. *Homo Sacer: Sovereign Power and Bare Life*. Trans. Heller-Roazen. Stanford: Stanford University Press, 1998.
Alcoff, Linda Martin. "Epistemologies of Ignorance: Three Types." In Shannon Sullivan and Nancy Tuana, eds., *Race and Epistemologies of Ignorance*. Albany: SUNY Press, 2007.
Alcoff, Linda Martin. "Epistemic Identities." *Episteme* 7, no. 2, 2010.
Alcoff, Linda Martin. *The Future of Whiteness*. Cambridge, UK: Polity Press, 2015.
Allen, Ernest Jr. "On the Reading of Riddles: Rethinking Du Boisian 'Double Consciousness.'" In Lewis R. Gordon, ed., *Existence in Black: An Anthology of Black Existential Philosophy*. New York: Routledge, 1997.
Anderson, Elizabeth. "Epistemic Justice as a Virtue of Social Institutions." *Social Epistemology* 26, no. 2, 2012.
Angell, Marcia. *The Truth about Drug Companies*. New York: Random House, 2005.
Arendt, Hannah. *The Human Condition*. Chicago: University of Chicago Press, 1928.
Aristotle. *Nichomachean Ethics*, trans. J. E.C. Weldon, London: Macmillan, 1927.
Axtell, Guy. "Expanding Epistemology: A Responsibilist Approach." *Philosophical Papers* 37, no. 1, 2008.
Babbitt, Susan. "Philosophy's Whiteness and the Loss of Wisdom." In George Yancy, ed., *The Center Must Not Hold; White women Philosophers on the Whiteness of Philosophy*. Lexington Books, 2010.
Baier, Annette. "Cartesian Persons." In her *Postures of the Mind: Essays on Mind and Morals*. Minneapolis: University of Minnesota Press, 1985.
Bailey Alison. "Strategic Ignorance." In Shannon Sullivan and Nancy Tuana, eds., *Race and Epistemologies of Ignorance*. Albany: SUNY Press, 2001.
Bailey, Alison. "White Talk as a Barrier to Understanding the Problem of Whiteness." In George Yancy, ed., *White Self-Criticality Beyond Anti-Racism: What Is It Like to Be a White Problem?* Lanham Maryland: Lexington Books, 2015.

Barad, Karen. *Meeting the Universe Halfway: Quantum Physics and the Entanglement of Matter and Meaning*. Durham, NC: Duke University Press, 2007.
Bartky, Sandra. "Sympathy and Solidarity: On a Tightrope with Scheler." In Diana Meyers, ed., *Feminists Rethink the Self*. Boulder, CO: Westview Press, 1997.
Beever, Jonathan, and Nicholae Morar. 'Bioethics and the Challenge of the Ecological Individual," in *Environmental Philosophy* 13, no. 2, 2016.
Benzaquen, Adriana. *Encounters with Wild Children: Temptation and Disappointment in the Study of Human Nature*. Montreal: McGill-Queen's University Press, 2006.
Biber, Eric. "Which Science? Whose Science?" How Scientific Disciplines Can Shape Environmental Law,." *The University of Chicago Law Review* 79, 2: Spring 2012.
Blum, Lawrence. "Moral Perception and Particularity." *Ethics* 101, no. 4, July 1991.
Brady and Pritchard, eds, *Moral and Epistemic Virtues*, special issue of *Metaphilosophy* 34, nos. 1–2 January 2003, 1–227.
Brown, Theodore L. *Making Truth: Metaphor in Science*. Urbana: University of Illinois Press.
Bunting, Madeline. "Two years on, Katine offers much to celebrate and much to feel frustrated about." *The Guardian*. Sunday, November 1, 2009.
Butler, Judith. *Bodies That Matter: On the Discursive Limits of Sex*. New York: Routledge, 1993.
Castoriadis, Cornelius. *Philosophy, Politics, Autonomy: Essays in Political Philosophy*, ed., David Ames. New York: Oxford University Press, 1991.
Castoriadis, Cornelius. "Radical Imagination and the Social Instituting Imaginary." In Gillian Robinson and John Rundell, eds., *Rethinking Imagination: Culture and Creativity*. London: Routledge, 1994.
Castoriadis, Cornelius. *The Imaginary Institution of Society*, trans. Kathleen Blaney. Cambridge: MIT Press, 1998.
Cavarero, Adriana. *Relating Narratives: Storytelling and Selfhood*. London: Routledge, 2000.
Cavarero, Adriana. *For More than One Voice: Toward a Philosophy of Vocal Expression*, trans, Paul A. Kotman. Stanford: Stanford University Press, 2005.
Cavarero, Adriana. *Inclinations: A Critique of Rectitude*, trans Amanda Minervini and Adam Sitze. Stanford: Stanford University Press, 2016.
Cheney, Jim. "Postmodern Environmental Ethics: Ethics as Bioregional Narrative." *Environmental Ethics* 11, no. 2 1989.
Coady, C.A.J. *Testimony: A Philosophical Study*. Oxford: Oxford University Press, 1992.
Cockburn, Cynthia. *The Space between Us: Negotiating Gender and Identities in Conflict*. London: Zed Books, 1998.
Code, Lorraine. "Is the Sex of the Knower Epistemologically Significant?" *Metaphilosophy* 12, nos. 3/4 July/October 1981.

Code, Lorraine. *Epistemic Responsibility*. Hanover, NH: University Press of New England, 1987.
Code, Lorraine. *Rhetorical Spaces: Essays on (Gendered) Locations*. New York: Routledge, 1995.
Code, Lorraine. *What Can She Know? Feminist Theory and the Construction of Knowledge*. Ithaca: Cornell University Press, 1991.
Code, Lorraine. "How to Think Globally: Stretching the Limits of Imagination." *Hypatia: A Journal of Feminist Philosophy* 13, no. 2 1998. Reprint in Uma Narayan & Sandra Harding, eds., *Decentering the Center": Philosophy for a Multicultural, Postcolonial, and Feminist World*. Bloomington: Indiana University Press, 2000.
Code, Lorraine. *Ecological Thinking: The Politics of Epistemic Location*. New York: Oxford University Press, 2006.
Code, Lorraine. "Advocacy, Negotiation, and the Politics of Unknowing." *Southern Journal of Philosophy* 46, no. 1, 2008.
Code, Lorraine. "Particularity, Epistemic Responsibility, and the Ecological Imaginary." *Philosophy of Education Archive* 2010, 23–34.
Code, Lorraine. "Testimony, Advocacy, Ignorance: Thinking Ecologically about Social Knowledge." In Allan Miller, Adrian Haddock, and Duncan Pritchard, eds., *Social Epistemology*. Oxford: Oxford University Press, 2010, 29–50.
Code, Lorraine. "'They Treated Him Well': Fact, Fiction, and the Politics of Knowledge" In Heidi Grasswick, ed., *Feminist Epistemology and Philosophy of Science: Power in Knowledge*. Dordrecht: Springer, 2011.
Code, Lorraine. "Virtue, Reason, and Wisdom." In Stan Van Hooft, ed., *Handbook of Virtue Ethics*, New York: Routledge, 2013, 188–99.
Code, Lorraine. "Ecological Subjectivities, Responsibilities, and Agency." In Anna Grear and Louis Kotze eds., *Research Handbook on Human Rights and the Environment*. Cheltenham: Edward Elgar, 2015, 46–60.
Code, Lorraine, "Epistemic Responsibility." In Ian James Kidd, Jose Medina, and Gaile Pohlhaus Jr., eds., *The Routledge Handbook of Epistemic Injustice*. London: Routledge 2017, 89–99.
Code, Lorraine. "The Tyranny of Certainty." In *Symposium: Canadian Journal of Continental Philosophy* 21, no. 1, Spring 2017, 206–18.
Collingwood. R. G., *An Essay on Metaphysics*. Oxford: Oxford University Press, 1939.
Collins, Gail. *As Texas Goes . . . : How the Lone Star State Hijacked the American Agenda*. New York: Norton, 2012.
Conley, Verena Andermatt, *Ecopolitics: The Environment in Poststructuralist Thought*. London: Routledge, 1997.
Cormier, Harvey. "Ever Not Quite: Unfinished Theories, Unfinished Societies, and Pragmatism." In Sullivan and Tuana, eds., *Race and Epistemologies of Ignorance*.

Craig, Edward. *Knowledge and the State of Nature: An Essay in Conceptual Synthesis.* Oxford: Clarendon Press, 1990.
Cuomo, Chris J. "Sexual Politics in Environmental Ethics: Impacts, Causes, Alternatives." In Stephen M. Gardner and Allen Thompson, eds., *The Oxford Handbook of Environmental Ethics*, Oxford University Press, 2017.
Dalmiya, Vrinda. "Why Should a Knower Care?" *Hypatia* 17, no. 1, January 2002.
Darwall, Stephen. *The Second-Person Standpoint.* Harvard: Harvard University Press, 2006.
Daston, Lorraine. "Objectivity and the Escape from Perspective." *Social Studies of Science* 22, no. 4, Nov. 1992, 597–618.
Daston, Lorraine. "Baconian Facts, Academic Civility, and the Prehistory of Objectivity." In Alan Megill, ed., *Rethinking Objectivity*. Durham, NC: Duke University Press, 1994, 37–63.
Daukas, Nancy. "Epistemic Trust and Social Location." *Episteme* 3, nos. 1–2 June 2006.
Deleuze, Gilles, and Felix Guattari. *What Is Philosophy?* Trans. Graham Burchell and Hugh Tomlinson. London: Verso Press, 1994.
Dixon, Mark, Director. 2017. *The Power of One Voice: A 50 Year Perspective on the Life of Rachel Carson.* Film.
Dotson, Kristie. "In Search of Tanzania" Are Effective Epistemic Practices Sufficient for Just Epistemic Practices?" *The Southern Journal of Philosophy* 46, no. 1, 2008.
Driver, Julia. "The Conflation of Moral and Epistemic Virtue." In Brady and Pritchard, eds., *Moral and Epistemic Virtues*. Oxford: Wiley-Blackwell, 2004.
Eflin, Juli. "Epistemic Presuppositions and Their Consequences." In Brady and Pritchard, eds., *Moral and Epistemic Virtues*. Oxford: Wiley-Blackwell, 2004.
Esquith, Stephen L. *The Political Responsibilities of Everyday Bystanders.* University Park: Pennsylvania University Press, 2010.
Figueroa, Robert M., and Gordon Waitt. "Climb: Restorative Justice, Environmental Heritage, and the Moral Terrains of Uluru-Kata National Park." *Environmental Philosophy* 7, no. 2 (Fall 2010), pp. 135–63.
Foucault, Michel. *Madness and Civilization: A History of Insanity in the Age of Reason*, trans., Richard Howard. New York: Vintage, 1965.
Foucault, Michel. "The Discourse on Language." In Michel Foucault, *The Archaeology of Knowledge and the Discourse on Language*. Trans. A.M. Sheridan Smith. New York: Pantheon Books, 1972.
Foucault, Michel. "What Is Enlightenment?" trans. Catherine Porter. In Paul Rabinow, ed., *The Foucault Reader*. New York: Pantheon Books, 1984.
Fricker, Elizabeth. "Testimony and Epistemic Autonomy." In Lackey and Sosa, eds., 2006.
Fricker, Miranda. *Epistemic Injustice: Power and the Ethics of Knowing.* Oxford: Oxford University Press, 2007.

Fricker, Miranda. "Epistemic Injustice and a Role for Virtue in the Politics of Knowing." In Brady and Pritchard, eds., *Moral and Epistemic Virtues*.

Frye, Marilyn. "In and Out of Harm's Way: Arrogance and Love." In her *The Politics of Reality: Essays in Feminist Theory*, Trumansburg, NY: The Crossing Press, 1983.

Gardiner, Stephen. *A Perfect Moral Storm: The Ethical Tragedy of Climate Change*. Oxford: Oxford University Press, 2011.

Gardiner, Stephen M. "Are We the Scum of the Earth? Climate Change, Geoengineering, and Humanity's Challenge." In Allen Thompson and Jeremy Bendik-Keymer, *Ethical Adaptation to Climate Change: Human Virtues of the Future*. Cambridge MA: MIT Press, 2012.

Gatens, Moira. "The 'Disciplined Imagination': Literature as Experimental Philosophy," *Australian Feminist Studies* 22, no. 52 March 2007, 25–34.

Gibbard, Allan. *Wise Choices, Apt Feelings: A Theory of Normative Judgment*. Cambridge, MA: Harvard University Press, 1990.

Gilligan, Carol. *In a Different Voice: Psychological Theory and Women's Development*. Cambridge, MA: Harvard University Press. 1993.

Glazebrook, Trish. "Heidegger and Ecofeminism." In Nancy J. Holland and Patricia Huntington,eds. *Feminist Interpretations of Martin Heidegger*. University Park: Pennsylvania State University Press, 2001.

Gordimer, Nadine. *July's People*. New York and London: Penguin Books, 1981.

Gordon, Lewis R., *Existentia Africana: Understanding Africana Existential Thought*. New York: Routledge, 2000.

Grasswick. Heidi, ed. *Feminist Epistemology and Philosophy of Science: Power in Knowledge*. New York: Springer, 2011.

Grasswick, Heidi, and Nancy A. McHugh, eds. *The Power of Example*. Albany: SUNY Press, forthcoming.

Hacking, Ian. "Language, Truth, and Reason." In Martin Hollis and Steven Lukes, eds., *Rationality and Relativism*. Cambridge MA: MIT Press, 1982.

Haddock, Adrian, Allan Millar, and Duncan Pritchard, eds. *Social Epistemology*. Oxford: Oxford University Press, 2010.

Haraway, Donna. *Modest_Witness@Second_Millenium. FemaleMan Meets_OncoMouse*. New York: Routledge, 1997.

Haraway, Donna. *Staying with the Trouble: Making Kin in the Chthulucene*. Durham, NC: Duke University Press, 2016.

Haraway, Donna J. "Situated Knowledges: The Science Question in Feminism and the Privilege of Partial Perspective." In her *Simians, Cyborgs, and Women: The Reinvention of Nature*. New York: Routledge, 1991.

Harding, Sandra. *The Science Question in Feminism*. Ithaca: Cornell University Press, 1986.

Hardwig, John. "Epistemic Dependence." *Journal of Philosophy*, 1985; and "The Role of Trust in Knowledge." *Journal of Philosophy*, 1991.

Haslanger, Sally. "What is natural and what is social?" *Feminist philosophers*: Posted: 22 Feb 2013 10.41 AM PST.
Hayward, Bronwyn. "The Social Handprint: Decentering the Politics of Sustainability After an Urban Disaster." In Peg Rawes, ed., *Relational Architectural Ecologies: Architecture, Nature and Subjectivity*. London: Routledge, 2013.
Hazlett, M. "'Woman vs. Man vs. Bugs': Gender and Popular Ecology in Early Reactions to Silent Spring," *Environmental History* 9, no. 4, Oct. 2004, 701–29.
Heidegger, Martin. *Being and Time*, trans. John Macquarrie and Edward Robinson. New York: Harper & Row, 1962.
Heidegger, Martin. "The Question Concerning Technology," trans. William Lovitt. In David Farrell Krell, ed., *Martin Heidegger: Basic Writings*. New York: Harper & Row, 1977.
Hemmings, Clare. "Affective solidarity: Feminist reflexivity and Political transformation." *Feminist Theory* 13, no. 2: 2012.
Hume, Mark. "Famous Medical Ethics Lecturer's Credentials Challenged in Euthanasia Case." Toronto: *The Globe and Mail*, November 6, 2011.
Hutton, Will. "Our planet needs us to fight for its survival," *The Guardian Weekly*, Sept. 2013.
Irigaray, Luce. *The Way of Love*. New York: Continuum Press, 2004.
Kruks, Sonia. *Retrieving Experience: Subjectivity and Recognition in Feminist Politics*. Ithaca: Cornell University Press, 2001.
Lacey, Hugh. Review of Kristin Shrader-Frechette, *Taking Action, Saving Lives*. In *Ethics* 118, no. 4, July 2008.
Lackey, Jennifer, and Ernest Sosa, eds. *The Epistemology of Testimony*. Oxford: Clarendon Press, 2006.
Latour, Bruno. "Irreduction." In Werner Callebaut, *Taking the Naturalist Turn; or, How Real Philosophy of Science is Done*. Chicago: University of Chicago Press, 1993.
Latour, Bruno. *The Politics of Nature; How to Bring the Sciences into Democracy*. Trans. Catherine Porter. Cambridge: Harvard University Press, 2004.
Le Doeuff, Michèle. *The Sex of Knowing*, trans. Kathryn Hamer and Lorraine Code. New York: Routledge, 2003.
Lloyd, Genevieve. *The Man of Reason: "Male" and "Female" in Western Philosophy*, 2nd ed. London: Routledge, 1994.
Lloyd, Genevieve. *Providence Lost*. Cambridge, MA: Harvard University Press, 2008.
Longino, Helen. *The Fate of Knowledge*. Princeton: Princeton University Press, 2002.
Lugones, Maria. "Playfulness, 'World'-Travelling and Loving Perception," *Hypatia: A Journal of Feminist Philosophy* 2, no. 2 1987, 3–19.

Lytle, Mark Hamilton. *The Gentle Subversive: Rachel Carson, Silent Spring, and the Rise of the Environmental Movement*. New York: Oxford University Press, 2007.
MacIntyre, Alasdair, *After Virtue: A Study on Moral Theory* University of Notre Dame Press, 1984.
Mackenzie, Catriona, and Natalie Stoljar, eds., *Relational Autonomy: Feminist Perspectives on Autonomy, Agency, and the Social Self*. New York: Oxford University Press, 2000.
Malterud, Kirsti, Lucy Candib, and Lorraine Code. "Responsible and responsive knowing in medical diagnosis." NORA 12:1, 2004.
Medina, José. "The Relevance of Credibility Excess in a Proportional View of Epistemic Dependence." *Social Epistemology: A Journal of Knowledge, Culture and Policy"* 25, no. 1, 2011, 15–35.
Medina, José. *The Epistemology of Resistance: Gender and Racial Oppression, Epistemic Injustice, and Resistant Imaginations*. Oxford: Oxford University Press, 2012.
Melo-Martin, Immaculada de, and Kristin Intemann. *The Fight against Doubt: How to Bridge the Gap between Scientists and the Public*. Oxford: Oxford University Press, 2018.
Messing, Karen. *One-Eyed Science: Occupational Health and Women Workers*. Philadelphia: Temple University Press, 1998.
Michaels, David. *Doubt Is Their Product: How Industry's Assault on Science Threatens Your Health*. New York: Oxford University Press, 2008.
Michaels. David. "Manufactured Uncertainty: Contested Science and the Protection of the Public's Health and Environment." In Proctor and Schiebinger, eds., *Agnotology*.
Millar, Allan, Adrian Haddock, and Duncan Pritchard, eds. *Social Epistemology*. New York: Oxford University Press, 2010.
Mills, Charles. "'Ideal Theory' as Ideology." *Hypatia: A Journal of Feminist Philosophy* 20, no. 3, 2005.
Misak, Cheryl. "Experience, Narrative, and Ethical Deliberation." *Ethics* 118, July 2008, 614–32.
Mohanty, Chandra Talpade. "Women Workers and Capitalist Scripts: Ideologies of Domination, Common Interests, and the Politics of Solidarity." In *Feminist Genealogies, Colonial Legacies, Democratic Futures*, M. Jacqui Alexander and Chandra Talpade Mohanty, eds. New York: Routledge, 1997.
Moran, Richard. "Getting Told and Being Believed." In Lackey and Sosa, eds., *The Epistemology of Testimony*, 2006, 272–306.
Narayan, Uma, and Sandra Harding, eds. *Decentering the Center: Philosophy for a Multicultural, Postcolonial, and Feminist World*. Bloomington: Indiana University Press, 2000.

Nelson, Julie A. "Is Dismissing the Precautionary Principle the Manly Thing to Do? Gender and the Economics of Environmental Protection." *Ethics and the Environment* 20, no. 1, March 2015, 99–122.

Nzegwu, Nkiru. *Family Matters: Feminist Concepts in African Philosophy of Culture.* Albany: SUNY Press, 2006.

Noddings, Nel. *Caring: A Feminine Approach to Ethics and Moral Education.* Berkeley: University of California Press, 1984.

Oreskes, Naomi, and Erik M. Conway. *Merchants of Doubt: How a Handful of Scientists Obscured the Truth on Issues from Tobacco Smoke to Global Warming.* New York: Bloomsbury Press, 2010.

Parker, Wendy S. "Environmental Science: Empirical Claims in Environmental Ethics." *The Oxford Handbook of Environmental Ethics*, Stephen M. Gardiner & Allen Thompson, eds., New York: Oxford University Press, 2017.

Passerin d'Entreves, Maurizio. "Hannah Arendt." In *The Stanford Encyclopedia of Philosophy* Winter 2013, ed. Edward N. Zalta.

Plumwood, Val. *Feminism and the Mastery of Nature.* London: Routledge, 1993.

Plumwood, Val. *Environmental Culture: The Ecological Crisis of Reason.* London: Routledge, 2001.

Plumwood, Val. "The Politics of Reason: Toward a Feminist Logic." In Rachel Falmagne and Marjorie Hass, eds., *Representing Reason: Feminist Theory and Formal Logic.* Lanham, MD Rowman & Littlefield, 2002.

Popper, Karl. *The Open Society and Its Enemies.* London: Routledge, 1945.

Popper, Karl. "Epistemology without a Knowing Subject." In his *Objective Knowledge: An Evolutionary Approach*, New York: Oxford University Press, 1979.

Powys Whyte, and Robert P. Crease, "Trust, Expertise, and the Philosophy of Science." *Synthese* 177, no. 3, 2010.

Proctor, Robert N., and Londa Schiebinger, eds. *Agnotology: The Making and Unmaking of Ignorance.* Stanford: Stanford University Press, 2008.

Rawes, Peg, ed. *Relational Architectural Ecologies: Architecture, Nature and Subjectivity.* London: Routledge, 2013.

Resznityk, Andrew. "Eyes through Oil." Unpublished PhD Dissertation, McMaster University, Canada.

Rich, Adrienne. *Blood, Bread, and Poetry: Selected Prose 1979–1985.* New York: Norton, 1994.

Rolin, Kristina. "Gender and Trust in Science." *Hypatia: A Journal of Feminist Philosophy* 17, no. 4, 2002.

Scott, Joan Wallach, ed. *Women's Studies on the Edge.* Durham, NC: Duke University Press, 2008.

Scheman, Naomi. "Empowering Canaries: Sustainability, Vulnerability, and the Ethics of Epistemology." *International Journal of Feminist Approaches to Bioethics* 7, no. 1 (Spring 2014).

Seager, Joni. "Rachel Carson Died of Breast Cancer: The Coming of Age of Feminist Environmentalism." *Signs: Journal of Women in Culture and Society* 28, no. 3 (2003) 945–72.

Seager, Joni. "Death by Degrees: Taking a Feminist Hard look at the 2(degrees) Climate Policy." *Kvinder, Kon & Forskning* 3–4, 2009, at p. 18.

Shotwell, Alexis. "Appropriate Subjects: Whiteness and the Discipline of Philosophy." In George Yancy, ed., The Center Must Not Hold, 117–130.

Shrader-Frechette, Kristin. *Environmental Justice: Creating Equality, Reclaiming Democracy*. New York: Oxford University Press, 2002.

Shrader-Frechette, Kristin. *Taking Action, Saving Lives: Our Duties to Protect Environmental and Public Health*. New York: Oxford University Press, 2007.

Sideris, Lisa H., and Kathleen Dean Moore, eds. *Rachel Carson: Legacy and Challenge*. Albany: SUNY Press, 2008.

Sim, Stuart. *Empires of Belief: Why We Need More Scepticism and Doubt in the Twenty-First Century*. Edinburgh: Edinburgh University Press, 2006.

Smith, Barbara Herrnstein. "The Unquiet Judge: Activism without Objectivism in Law and Politics." In Allan Megill, ed., *Rethinking Objectivity*. Durham, NC: Duke University Press, 1994.

Smith, Michael B. "'Silence, Miss Carson!' Science, Gender, and the Reception of *Silent Spring*." *Feminist Studies* 27, no. 3 (Autumn 2001), 733–52.

Smith, Mick. *An Ethics of Place: Radical Ecology, Postmodernity, and Social Theory*. Albany: SUNY Press, 2001.

Stone, Alison. *Luce Irigaray and the Philosophy of Sexual Difference*. New York: Cambridge University Press, 2006.

Sullivan, Shannon. *Revealing Whiteness: The Unconscious Habits of Racial Privilege*. Bloomington: Indiana University Press, 2006.

Sullivan, Shannon. "White Ignorance and Colonial Oppression: Or, Why I Know So Little about Puerto Rico." In Shannon Sullivan and Nancy Tuana, eds., *Race and Epistemologies of Ignorance*. Albany: SUNY Press, 2007.

Sullivan, Shannon. *Against Ecological Sovereignty: Ethics, Biopolitics, and Saving the Natural World*. Minneapolis: University of Minnesota Press, 2011.

Sullivan, Shannon. *Good White People: The Problem with Middle-Class White Anti-Racism*. Albany: SUNY Press, 2014.

Tolstoy, Leo. *Anna Karenina*, trans. R. Pevear and L. Volokhonsky. London: Penguin, 2004.

Troop, William M. "Environmental Virtues and the Aims of Restoration." In Thompson and Bendik-Keymer, *Ethical Adaptation to Climate Change*.

Vogel, Steven. "Nature as Origin and Difference. On Environmental Philosophy and Continental Thought. *Philosophy Today*. SPEP Supplement 1998, 169–81.

Voltaire. *Candide*, trans. Daniel Gordon. Boston: St. Martin's Press, 1999.

Welbourne, Michael. *The Community of Knowledge*. Aberdeen: Aberdeen University Press, 1986.
Welbourne, Michael. *Knowledge* Montreal: McGill-Queen's University Press, 2001.
Whitford, Margaret. *Luce Irigaray: Philosophy in the Feminine*. London: Routledge, 1991.
Winch, Peter. "Nature and Convention." In *Ethics and Action*. London: Routledge and Kegan Paul, 1972.
Wittgenstein, Ludwig. *On Certainty*, ed. G. E. M. Anscombe and G. H. von Wright, trans. Denis Paul and G. E. M. Anscombe. Oxford: Blackwell, 1968.
Wittgenstein, Ludwig. *Philosophical Investigations*, trans. G.E.M. Anscombe. Oxford: Basil Blackwell, 1968,
Woolf, Virginia. *A Room of One's Own*. London, UK, Granada Books, 1977 (originally published 1929).
Yancy, George, ed. *The Center Must Not Hold: White Women Philosophers on the Whiteness of Philosophy*. Lanham, MD: Lexington Books, 2010.
Young, Iris Marion. *Inclusion and Democracy*. Oxford: Oxford University Press, 2000.
Young, Iris Marion. "Activist Challenges to Deliberative Democracy." *Political Theory* 29, no. 5, October 2001.
Ziarek, Krzystof. *Inflected Language: Toward a Hermeneutics of Nearness: Heidegger, Levinas, Stevens, Celan*. Albany: SUNY Press, 1994.

Index

A Room of One's Own (Virginia Woolf), 63
activism: from democratic deliberation, 119, 133; on climate change, 47, 67–68; and epistemic identity, 78, 84; knowing responsibly for, 56; risks of epistemic violence, 183
Adams, Douglas, 23
advocacy/advocates: assessment/positionality of advocates, 106, 110; as biased, non-credible knowledge, 77, 105–106; as communal, collaborative, 78; feminization of, 96–97; as irresponsible/harmful, 108, 133; of for-profit corporations, 116; as responsible epistemic practice, 61, 104, 113, 133
affluent white Northern-Western world: cognition impaired by bias/distortions, 166; and epistemic injustice, 165; notions of science, 67; philosophers/activists, 182–83; protection of status quo, 48; radical re-imagining of/by, 83–84; social imaginary of, 2–4, 10, 14–15, 65, 157–59, 172–73; threatened by climate science, 73–74
Africa/African women, 150–52

agency: assumptions about, 96, 101, 113, 169; of the autonomous individual, 41–42, 68, 124; in feminist/postcolonial thought, 31; of responsible knowers, 191, 206; transformed concept of, 59, 78, 111, 211
agnotology, 60, 67
Agnotology: The Making and Unmaking of Ignorance (Robert N. Proctor and Londa Schiebinger), 76, 82, 192, 195
Alcoff, Linda, 163, 169
Alexandrovich (*Anna Karenina*), 122–24
Allen, Ernest, Jr., 84–85
American exceptionalism, 70–71
An Ethics of Place (Mick Smith), 83
analogical thinking, 187, 189–90
analytic philosophy, 4, 11
anecdotes—as method, 120, 121
Angell, Marcia, 100–13; credentials, 100–101; credentials challenged, 103, 107–108, 110–11; as expert who cares, 105, 110
Anglo-American mainstream epistemology/philosophy: affect-free, 20, 123–24, 134; "average man" concept, 186; knowledge without a knower, 17, 113,

Anglo-American mainstream epistemology/philosophy *(continued)* 122–24, 126, 128; opacity unaddressed, 39–40; Othering practices of, 160; and particularity/uniqueness, 31–32, 144–45; philosophical certainty and truth, 8; place, concept of, 28, 64; radical individualism/entitlement, 3, 14, 58; tenets of, 143, 145, 164, 185–86; validity, conditions of, 29, 35, 190, 210. *See also* autonomous, unsituated (male) knower; empiricism; epistemic responsibility; instituted social imaginary; positivism; "view from nowhere"; individual/individualism

Anna Karenina (Leo Tolstoy), 122–25, 127–29, 132, 137

anti-ecological thought/action, 76

antiracism/critical race theories, 11, 14, 52, 53, 55, 56. *See also* critical theories of knowledge; gender-race-class specificities

antivaccination debates, 117

Arendt, Hannah, 17, 23–24, 33–34, 37, 67, 146

Aristotle, 23, 201, 203, 210, 212

authorities; authoritative knowers. *See* experts/expertise

autonomous, unsituated (male) knower: "the average man" research, 186; as caricature of modern humanity, 84; and detached objectivity, 41, 70, 79, 110; man of reason, 10, 159, 206, 208; one-dimensional, homo-economicus, 83; unmarked masculine norm, 80–81, 201; veil of neutrality, 67–68, 71, 79. *See also* "view from nowhere"; individual/individualism

Axtell, Guy, 129

"Ayers Rock" (Uluru), 13, 58

Babbitt, Susan, 12, 57, 204, 208–209, 212, 214

Bad Science: A Resource Book, 74–76, 87

Baier, Annette, 33, 146, 204, 212

Bailey, Alison, 39, 192–93, 197, 198

Balog, James, 47, 48

Barad, Karen, 113, 134, 185, 201

"bare life" (Agamben), 21

Bartky, Sandra, 16–17, 19, 209

Beauvoir, Simone de, 34, 95, 217

Beevor, Jonathan, 42, 43

Biber, Eric, 86

"Bioethics and the Challenge of the Ecological Individual" (Beevor and Morar), 42

biomedical research/thought, 43, 45–46

Blackness Visible (Charles Mills), 209

Blum, Lawrence, 32–33, 145–46

brain cancer, 121

breast cancer, 88, 90

Brown, Donald, 178–80

Brown, Theodore, 44

Brown, Wendy, 31

"buffered self" (Lloyd), 34, 39, 72, 147, 175

Bunting, Madeleine, 150–54, 156, 158, 174

Butler, Judith, 48–49

bystanders, 215–16

Camus, Albert, 217

Candide (Voltaire), 15

capitalism, 15, 70, 72–73, 83, 96

care/affect: as compromising "objectivity," 103, 122, 134; cost of its absence, 123; difficulty of incorporating, 126–27; as double-edged for women, 111; for nature/earth, 20; in producing knowledge, 12, 111, 122; as responsible epistemic practice, 57, 99, 114, 134, 214, 216; and techno-science, 60

Carson, Rachel, 85–97; critiques of, 88, 93–94, 93n55; death of, 87–88; as ecological subject, 85–86, 97; female identity noted, 68, 88–89; feminist inquiry of, 91; humility of, 204, 213, 214; methodology of, 85–86, 90–92, 94, 184; precautionary principle of, 180–81

Castoriadis, Cornelius: critique of orthodox epistemology, 164; instituted imaginary, 4, 14–15, 59, 170, 172–73; instituting imaginary, 101, 160

categorization, 57, 65

Cavarero, Adriana: natality/mortality, 33–34; particularity/uniqueness, 29–31, 41, 142–44, 146; politics and speech, 36–37; relationships/relationality, 32–34, 146–49

The Center Must Not Hold (George Yancy), 12, 57

certainty of knowledge, 8, 44, 123–25, 177, 180. See also uncertainty/doubt

Chasing Ice (2012 film—James Balog), 47–49

Cheney, Jim, 194

Christchurch, New Zealand, 17

citizens and knowledge production. See democratic deliberative inquiry

climate change: and complacency, 54; as complex phenomenon, 8; evidence of, 47, 48; ignorance of, 216

climate change science: as alarmist, 73–74; confusing for non-experts, 77; contested knowledge of, 54; fighting science with science, 68; weight to "both sides," 70–71

climate change skepticism: and the autonomous subject, 68; and cynicism, 67, 69; discriminatory, 94; versus ecological thought, 66–67; features of, 69, 173; freedom jeopardized, 67, 70, 73; harms of, 55, 56; introduction to, 1, 4–5; and manufactured doubt/ignorance, 46, 54, 56, 67, 178–80; and "normatively inappropriate dissent," 70; response to *Chasing Ice*, 47

Cockburn, Cynthia, 39

"cognitive sanctuary" (Fricker), 171

cohabitation, ideals of, 58

collectives versus individuals, 24, 173

colonialism, 82, 157–58

communities/ecosystems, 43, 206

Competitive Enterprise Institute, 93

complacency, 56

Conley, Verena, 85–86

consciousness-raising practices: on ecological living, 17, 66; feminist/anti-racist, 14, 53, 171, 207; "who we are," 3–4, 59

Conway, Eric, 7, 54, 59–60, 67, 70–76, 79, 193

corporate interests, 77, 119. See also pharmaceutical industry; tobacco industry

Craig, Edward, 35–36, 43, 147–50, 155, 171

critical theories of knowledge, 56, 180, 196, 200–201

cultural relativism, 183

Cuomo, Chris, 82, 83

"cup on table" as truth, 38, 82, 103, 105, 107, 134, 200

Dalmiya, Vrinda, 134, 216

Daston, Lorraine, 105–106, 118
Daukas, Nancy, 132, 133
David Copperfield (Charles Dickens), 212
DDT (dichloro-diphenyl-trichloroethane), 88–90, 92
Decentering the Center (Narayan/Harding, eds.), 191
Deepwater Horizon oil rig, 49
Deleuze, Gilles, 195–96
democratic deliberative inquiry: care and time required, 118–19; citizens and knowledge production, 18, 120; of responsible epistemic community, 77, 108; specificity and complexity addressed by, 107; and unsettling of status quo, 112
Descartes, René, 206
Dickens, Charles, 212
difference/diversity, 30, 141–42, 175, 184, 209
distrust, 56, 69
Dotson, Kristie, 154–56, 189
"double consciousness" (Du Bois), 84–85
doubt. *See* uncertainty/doubt
"doxastic shock," 152, 161
Driver, Julia, 214
Du Bois, W. E. B., 84–85

"ecological" as concept, 1–2
ecological harms, 58–59
ecological imaginary, 66, 73, 86, 153, 158–60
ecological subjectivity, 13, 64–66, 86–87, 160
Ecological Thinking (Lorraine Code): advocacy, 104, 106; autonomous individual, 42–43; "bio-regional narratives," 194; ecological subjectivity, 86; instituted imaginary, 101; mastery, 65, 159; responsibility; ideal cohabitation, 10, 30, 55–56, 113, 211; specificity, 38, 186
ecological thought: activist/instituting practices of, 67–68; and ambiguity, 95; versus climate change skepticism, 66; introduction to, 5, 211; longitudinal/horizontal lens of, 90; social imaginary of, 65–66; as socio-political, 95; and subjectivity, 9, 81, 85–86; transformative power of, 159–60
education/educational institutions, 24, 53
educators' responsibilities, 4, 12, 15, 50–51, 57, 59
Eflin, Juli, 205–206
Ellison, Ralph, 209
empathy, 216–17
Empires of Belief (Stuart Sim), 199
empiricism: "empirical simples" (cup on table), 38, 82, 103, 105, 107, 134, 200; "knowledge by description" (Russell), 153; limitations of, 90, 125; seeing is believing, 48–49; "statistics have shown," 116
Enlightenment, 15, 65, 110, 159
entitlement, 2, 9–10, 13, 57, 58
"environmental" as concept, 1
Environmental Protection Agency (EPA), 72–73, 75–76
Environmental Tobacco Smoke (ETS), 74
epistemic agency, 141
epistemic communities, 96, 119
epistemic distrust, 7
epistemic exclusion, 132–33
epistemic friction, 155–56, 160, 166, 189, 216
epistemic identity, 78–97, 163, 169
epistemic impairment, 165–66

epistemic individualism. *See* individual/individualism

epistemic injustice: and account of Tanzanians, 156; and "doxastic shock," 152; emancipatory potential of addressing, 166–67; Fricker's work, 37–38, 105n9; and politics of knowledge, 214; and the social imaginary, 164–66; treatment of Angell, 105

Epistemic Injustice: Power and the Ethics of Knowing (Miranda Fricker), 37–38, 147

epistemic interdependence (Hardwig), 131, 207

epistemic legitimacy (whose knowing counts), 13

epistemic literacy, 153

epistemic objectivity, 211

epistemic principle of charity, 131

epistemic responsibility: affective engagement, 189; analogical thinking, 187, 189–90; care/affect, 212; and climate change skepticism, 46; decentring traditional epistemologies, 191; defined, 2; emergence of concept, 101–102; and "epistemic identity" of inquirers, 170, 173; as ethical/virtuous, 200; features of, 14, 27–28; and humility, 184–85, 188, 190, 212; importance of "place," 28; interactive/collaborative methods, 184–85, 212; knowing others as model for, 134; lapses in, 154–56; limits of knowledge, 24, 175; myriad, delicate and crucial, 158; narratives as method, 194–95, 198; ontological aspects, 14; and pedagogy, 23–24, 96; and the racialized imaginary, 155; radical rethinking, 3–4, 14, 20, 212; redistributed epistemic authority, 186; renewed interest in, 46; in representing others, 150; scope of, 200; and situated knowers, 102; time required, 117–18, 184–86, 188, 190

Epistemic Responsibility (Lorraine Code), 27–28, 30, 102, 171

epistemic violence/violations, 40, 133, 183, 208, 209

epistemic vulnerability, 30

epistemic/cultural imperialism, 182–83, 190

epistemological vulnerability, 89

"epistemology of mastery" (Plumwood). *See* mastery

The Epistemology of Resistance (José Medina), 147

Esquith, Stephen L., 215, 217

ethics and epistemology, 23, 48, 119, 172, 211

euthanasia case (B.C. Supreme Court), 100–101, 103, 109

evidence as anecdotal, 50

examples, power of, 215, 216

experience as knowledge, 40

experts/expertise: and advocacy, 103; conflicting testimonies of, 68; credibility/trustworthiness of, 53, 94, 100–101, 108–109; deference to, 194; as knowers, 61; public trust in, 52

exposure of the self, 34, 147, 148

Fanon, Frantz, 209

female genital mutilation, 183

feminist epistemology, 5, 14, 160, 173; and antiracist theories, 11, 56, 191; asks "whose knowledge?," 146; critique of neutrality, 72; critique of the autonomous unmarked subject, 41–42, 54, feminist epistemology (continued) 80, 129; dialogic approach to knowledge, 180; eco-feminism, 9, 20; and opacity, 39–40; politics and ethics of knowledge, 48, 56; radical feminism, 22; rethinks ontological assumptions, 5, 14, 160, 173; social transformation/social justice, 3, 5, 11, 52, 84, 161, 196; subjectivity and knowledge, 80–82; Western feminism and white ignorance, 198. See also care/affect; consciousness-raising practices; particularity/uniqueness; postcolonial insights/theories; "situated knowledge" (Haraway); social epistemology; testimony

Ferrari, Michelle, 86

Ferro Chemical Plant, Chicago, 121

The Fight Against Doubt (Melo-Martin and Intemann), 69

Figueroa, Robert M., 13, 58

For More than One Voice (Adriana Cavarero), 45

Foucault, Michel: conditions of possibility, 168; "critical ontology of ourselves," 21; "dare to know," 202, 214; power/knowledge system, 52, 164; social imaginary, concept of, 171–72; "within the true," 145

Fox News, 50

freedom/Liberty: as American/white capitalist value, 71–72; jeopardized by science/regulation, 67, 73; and political inequality, 79; refusals to change behaviour, 55; and skepticism, 2, 70; and the unmarked male subject, 80, 81; "whose freedom are we talking about?," 79, 81

Fricker, Miranda: addressive knowledge-production, 147; epistemic injustice, 35, 105; epistemic sensibility, 205, 209; "the imagination," 167, 171; neutrality, 169; testimonial/hermeneutic injustice, 37–38

Frye, Marilyn, 192

Gardiner, Stephen, 22–23

Gatens, Moira, 154, 161

gender and global injustice, 82

gender-race-class specificities: erased in climate change skepticism, 68, 72; gender-cognizant standpoint, 82; glossed over, 31; in identifying who "we" are, 55; knowing well through difference, 81, 183, 207; and place, 64; and power, 54; and vulnerability, 51, 115. See also critical theories of knowledge; feminist epistemology; Othered people/subjects

geographies in knowledge-production, 195, 198

Gilligan, Carol, 97

Glazebrook, Trish, 20

global warming, 71, 216

Gordimer, Nadine, 196

"gravel," 63–64

greenhouse gas emissions, 60, 179

Guattari, Felix, 195–96

Haraway, Donna: assessing advocates, 106; "the god trick"/view from nowhere, 65, 107; "modest witness," 204, 213, 214; "situated knowledge," 28, 114, 148, 170, 177–78, 203

Harding, Sandra, 114, 131–32, 172

Hardwig, John, 122, 125–32, 135–38

harms, 1–2, 60, 82

Haslanger, Sally, 57

Hayward, Bronwyn, 17–18

health: effects of chemicals on, 121, 151; studies in Tanzania, 183–85, 188; workplace health, 186
Heidegger, Martin, 19–21, 34, 48, 147, 151–52
Hemmings, Claire, 215–17
hermeneutic humility, 183–90
hermeneutic injustice, 164, 166, 170
homophobia, 206
human nature/subjectivity/agency: assumptions about, 4, 16, 70, 79, 86; diverse enactments of, 141; human sameness, 2, 57, 79, 86, 112, 141, 182, 196–97; renewed conceptions of, 59
"human rights and the environment," 4, 56
human suffering, 60
Hume, Mark, 100
humility: as acknowledgement, 27–28; in advocacy, 77; to avoid "rush to judgement," 81; in epistemic practice, 99–100, 181–82, 184–85, 188, 190; and epistemic responsibility, 210; lack of, 185; modesty, 213–14; practiced by writers/scholars, 213; as restraint of self-interest, 24; and Uriah Heap, 212–14
Hutton, Will, 95
Hypatia article (Code), 182, 191, 192

identity politics, 11, 31
IDRC (International Development Research Centre), 182, 183–84, 186–88
Igbo women, Nigeria, 208
ignorance: and climate science, 193; counter-acting arrogance, 192; culpability for, 46–61; dangers of, 199; defined by Bailey, 197; effects/harms of, 44, 193; epistemologies of, 101; and limitations of knowledge, 199; production of, 56, 76; social ignorance, 47–48, 54; strategic, 191–200; white ignorance, 199. *See also* uncertainty/doubt
"ignorance/knowledge" (Sullivan), 25, 51, 157, 174, 193
"imaginary" as concept, 65n6
imagination, 167, 170, 171
incommensurability, 185
incredulity, 30, 56
Indigenous Australians, 58
individual/individualism: and climate change denial, 65; critiques of, 41–46, 109, 129; limits of individualistic knowledge, 24, 126–27; and positivism, 20; self-certainty, 58; versus "the social imaginary," 167; turn away from, 192; in Western philosophy/liberalism, 3, 9, 57, 83, 148. *See also* autonomous, unsituated (male) knower; "buffered self" (Lloyd); entitlement; "view from nowhere"
inquiry/knowledge production: adjudication of differences, 118; anecdotes versus facts, 92; and colonialism, 157–58; as cooperative; socio-political, 28; different modes of, 58; erasure of work that produces data, 112–13; estrangement/decentring/defamiliarization, 191–92; individualistic versus communal, 94; masculine/feminine approaches to, 80; multiple ways of knowing, 102, 126; new logics of, 192, 197–98; as place-based, 64; as possessing information, 16–17; power-infused, real world aspects, 109. *See also* democratic deliberative inquiry;

inquiry/knowledge production (*continued*)
empiricism; knowers/knowing subjects; "one-liners": facts as truth; "S knows that p"
instituted social imaginary: circumscribed imagination of, 104; cognitive aspects of, 169; and cultural imperialism, 190; defined by Castoriadis, 15; disruption/destabilization of, 53, 59, 83–84, 161–62, 174; effects of, 101, 150, 158–59, 164, 170; explanatory/social power of, 71, 165, 171; features of, 44, 65, 159, 172, 180, 188; as focus of epistemic inquiry, 163–64; global scope of, 183–84; incongruous with itself, 15, 101, 161, 170, 173; and injustice/irresponsibility, 160, 166, 206; political and ethical dimensions of, 167; of Popper and Tolstoy, 125; of Western society, 3, 4, 14, 59, 105, 107, 194. *See also* instituting imaginary; objectivity
instituted social-political-epistemic imaginary. *See* instituted social imaginary
instituting imaginary, 15, 101, 160–61
instrumentalism, 9, 9–10, 156
Intemann, Kristen, 69, 126
International Interdisciplinary Conference on Women (Australia, 1996), 182
international treaties and cautionary principle, 215
interviewing as method, 155–56
intra-active knowledge production (Barad), 113
Invisible Man (Ralph Ellison), 209

IPCC (Intergovernmental Panel on Climate Change), 7, 95
Irigaray, Luce, 64–65, 79

Janzen, Grace, 34
Johnson, Patricia, 34
July's People (Nadine Gordimer), 196

Kant, Immanuel, 142
Katine project, Uganda, 150–53, 158
King, Rodney, 48
knowers/knowing subjects: from detached to addressive, 35, 147, 149, 210, 212; not *visibly* political, 128; responsible knowers, 28, 44, 52, 108; as self-transforming/relational, 16–17; as situated, 80–82, 114, 178–79; as univocal/unsituated, 75, 147, 155–56; whose knowledge counts/is recognized, 30, 32, 82. *See also* autonomous, unsituated (male) knower; gender-race-class specificities; Othered people/subjects; "situated knowledge" (Haraway); women and inquiry
knowledge, limits of, 175
knowledge production. *See* inquiry/knowledge production
Knowledge and the State of Nature (Edward Craig), 147
knowledge claims, 102–103
knowledge/ignorance (Sullivan), 25, 51, 116, 174
Kruks, Sonia, 16

laboratory science, 44–45
Latour, Bruno, 112, 194, 195
Lawrence, Stephen, 168
Le Doeuff, Michèle, 163–64, 166–67, 172

linguistic shifts: third person to speakers/hearers, 35, 147, 210, 212
listening as epistemological method, 17–20, 34, 198
literature as knowledge, 196–97, 217
Lloyd, Genevieve, 10, 34, 147, 191, 207
Longino, Helen, 124–25, 129, 130–31, 135–36
longitudinal research, 90
Lugones, Maria, 39, 85, 184, 192, 195

Making Truth: Metaphor in Science (Theodore Brown), 44
man of reason, 10, 159, 206, 208
The Man of Reason: "Male" and "Female" in Western Philosophy (Genevieve Lloyd), 10, 191
manipulation, 54
manufactured uncertainty, 61, 163, 193–94
mastery, 73, 80, 83, 90, 95, 159, 160
Medina, José, 46, 48–50, 53, 147, 155, 165–66, 169–70
Melo-Martín, Immaculada de, 69, 126
Merchants of Doubt (Oreskes and Convway): analysis of public skepticism, 59–60, 67, 70; treatment of Carson's work in, 88, 88–90, 92, 94, 96–97; treatment of subjectivities in, 78–79, 81
Messing, Karen, 38, 186
Michaels, David, 76–77, 116, 193–94
Mills, Charles, 191, 199, 209
Mohanty, Chandra, 197–99
moral knowledge, 120
moral particularity, 32, 145
"Moral Perception and Particularity" (Lawrence Blum), 32, 145

Morar, Nicholas, 42, 43
mortality, 34
multiplicity: assumptions about, 32; and ecological subjectivity, 13, 42, 58, 192; in investigating social (in)justice, 11, 56; knowing well, 28, 29, 210

narratives: destabilization of established narratives, 161, 174, 199; epistemic legitimacy of, 13, 38, 120–21, 194–96; knowing well, 198; master narratives, 159; relations among people in, 33, 58, 146. *See also* stories/storytelling
natality, 146
Nation of Change, 21
Nausea (Jean-Paul Sartre), 217
Nelson, Julie, 214–15
neutrality, 167, 169, 172
New England Journal of Medicine, 100
Nigerian women, 208
"normatively inappropriate dissent" (NID), 69
"Notes for a Politics of Location" (Adrienne Rich), 22

objectification/instrumentalization, 149
objectivity: and advocacy, 120–35; care/interest untenable, 99, 107, 122; of detached/featureless knower, 11, 41, 105–106, 110, 123, 142; as dogma, 104; and mastery/manhood, 80; in social epistemology, 211; versus social imaginary, 163; "strong objectivity" (Harding), 114; value of, 105, 118, 123
O'Brien, Mary, 34
oil spills, 49

On Certainty (Ludwig Wittgenstein), 8
"one-liners": facts as truth: limitations of, 38, 109, 119–20, 120, 198, 216; and testimony, 35, 37, 40, 147–48, 148, 175; and who "we" are, 23
ontology: "ontology of the self," 12, 21, 57; reconception of, 4n3, 13, 58; of Western society, 9–10, 55, 165, 168
opacity, 175
The Open Society and Its Enemies (Karl Popper), 124
Oreskes, Naomi, 7, 54, 59–60, 67, 70–76, 79, 193
Othered people/subjects: care as double-edged, 111; effects of climate change on, 7, 11, 178; harm to/mastery over, 65, 159; in literature, 217; preserving world for, 11, 56, 200; privacy/opacity of, 174–75; respect for, 13, 18n28, 58, 108; responsible epistemology of, 11, 56, 80; and specificity/disaggregation, 115; treatment in hegemonic epistemology, 129, 145, 160, 172, 209. *See also* critical theories of knowledge; postcolonial insights/theories; race/racism

particularity/uniqueness: and abstraction, 40–41; in Caverero's work, 29–30; as distinct from individualism/positivism, 30–31, 144; of each human being, 142–43; and ecological thought, 66; feminization of, 142–43; may hide structural/systemic aspects, 168; as pedagogical resource/praxis, 61; as stymying "objectivity," 41, 61, 142; valuing of, 29

Pearson, Emily, 121
pedagogical practices, 4, 15–16, 24, 59, 174, 216
pharmaceutical industry, 193–94
philosophy defined, 195–96
photographs as "evidence," 49
physical sciences, 131
Plumwood, Val: call for ethical science, 48; and care/affect, 12; as feminist ecological philosopher, 9–19, 50; man as default norm, 81; mastery, epistemology of, 73, 159; women's roles in man's world, 79, 142
politics of knowledge, 51, 111–12, 214
The Political Responsibilities of Everyday Bystanders (Stephen L. Esquith), 215
Popper, Karl, 122, 124, 125
positivism, 27, 31, 180
postcolonial insights/theories, 14, 31, 84, 160
postpositivism, 44, 54, 104, 122, 163, 185
power, 51, 164–68, 172
precautionary principle, 179, 180–82, 214–15
Prigogine, Ilya, 85
private-interest science, 118, 120
Proctor, Robert, 193
public deliberations. *See* democratic deliberative inquiry
public trust in knowledge, 52–53, 119
Puerto Rico/Puerto Ricans, 157–59, 199

Race and Epistemologies of Ignorance (Sullivan/Tuana), 192
race/racism, 48, 60, 82, 154–56, 168, 206, 216

Rachel Carson (film, Ferrari), 86, 91
Reich, Robert, 21–22
Reidel, Charley, 49
"relating narratives" (Cavarero), 33
Relating Narratives: Storytelling and Selfhood (Adriana Cavarero), 29–30, 143–44
relationships/relationality, 9, 29, 33–34, 143–44, 146–47
research—public trust in, 52–53
Reszitnyk, Andrew, 48, 49
Retrieving Experience (Sonia Kruks), 16
Rich, Adrienne, 22
risk, 181
Rolin, Kristina, 132
Romney, Mitt, 178–80
Russell, Bertrand, 153

"S knows that p," 147–48, 153, 158, 195, 210, 212
Sartre, Jean-Paul, 205, 217
Scheman, Naomi, 44–45
The Science Question in Feminism (Sandra Harding), 131
science/scientific knowledge claims: certainty of knowlege, 178; in defense of status quo, 119; evaluation of, 74; versus facts, 69; hierarchies of, 92; masculinization of, 92; as neutral, objective, 67; resistance to scientific claims as situated, 75; "science has proved," 75, 102, 172, 173. *See also* climate change science
"scum of the earth," 22–24
Seager, Joni, 178–81, 184–85
"second persons," 146, 149, 204, 212
secularity/secularism, 12–13, 58
"seeing is believing," 48–49
Seitz, Frederick, 71
self-reflexivity, 114

Seller, Anne, 42–43
Serres, Michel, 85
sexism, 82, 206, 216
sexuate subjects, 65–66, 85, 96–97
She Came to Stay (Simone de Beauvoir), 217
Shotwell, Alexis, 13, 58, 209
Shrader-Frechette, Kristin, 113–22; affect as strengthening validity, 114, 121–22; care/affect in knowledge-production, 60, 111, 122, 123; methodology of, 114, 162; testimony, respect for, 122
silence, 175
Silent Spring (Rachel Carson), 93
Sim, Stuart, 199
"situated knowledge" (Haraway): absent from mainstream inquiry, 170, 177; and democratic deliberation, 18; as groundbreaking concept, 106, 148, 178; and place, 28; and social epistemology, 180, 203
skepticism. *See* climate change skepticism
Smith, Michael, 83, 84, 93
social change, 79, 137, 171. *See also* activism; advocacy/advocates; instituting imaginary
social cognitive ecology, 137–39
social epistemology: and ecology, 11; emergence/development of, 43, 101, 171, 180, 201; and epistemic identity, 163; negotiation/deliberation in, 210; reclaims the world, 210–11. *See also* epistemic responsibility; Hardwig, John; narratives; storytelling; testimony
social imaginary. *See* instituted social imaginary
social imagination (Fricker), 167, 171
social injustice, 169, 196

social justice: and ecological responsibilities, 54; and epistemic responsibility, 114–15; knowledge-based advocacy for, 119; practices of particularity, multiplicity, commonality, 56; as primary concern, 3, 11, 56; and understanding of others; empathy, 16
social knowledge, 129
social sciences, 131
solidarity, 217
Southern Journal of Philosophy, 154
specificity in knowledge production, 115, 141–42, 145, 184
spectator epistemology, 35–36, 90, 112, 147, 149
"standing reserve," earth as (Heidegger), 20–21
standpoint epistemology, 82, 85, 87. See also Harding, Sandra; "situated knowledge" (Haraway)
statistics—limitations of, 116
Stenger, Isabelle, 85
stereotypes, 38, 188
Stone, Alison, 65, 97
storytelling, 33, 120, 150, 153–54, 161–62, 174. See also narratives
The Stranger (Albert Camus), 217
strategic ignorance, 191–200
strategic skepticism, 54, 61, 174, 181–82, 199–200
"strong objectivity" (Harding), 114
subjectivity, 67, 68. See autonomous, unsituated (male) knower; gender-race-class specificities; "view from nowhere"; individual/individualism
Sullivan, Shannon: ignorance/knowledge, 25, 51, 116, 174; on secularity, 12, 58; white ignorance, 158–59, 192–93, 199

Taking Action, Saving Lives: Our Duties to Protect Environmental and Public Health (Kristin Shrader-Frechette), 113–16, 119–20
Tanzania/Tanzanian voices, 154–56, 183–89, 198
teachers, vulnerabilities of, 51–52
techno-science, 60, 162
testimony: addressive versus spectator, 21, 35, 147–48; in climate change debates, 67; constraints on comprehension, 167; in contexts of asymmetrical power, 60, 149; exposure/vulnerability/opacity, 35, 37, 40, 153, 175; as method of producing knowledge, 28, 83, 171, 212; reliability/credibility of, 129–30, 141; of scientific experts, 60–61; testimonial injustice, 38, 164, 166, 170; validity of/trust in, 38–39, 68, 92, 122, 148–49, 162. See also narratives; storytelling
textbooks, 50–51
thalidomide, 216
"think global, act local," 182–83, 191, 199, 200
Throop, William, 24
time needed for inquiry, 184–86, 188, 190
tobacco industry, 116, 193
tobacco use, 6, 60, 73, 75, 116
Tolstoy, Leo, 122–25, 127–37
The Truth About Drug Companies (Marcia Angell), 100
trust: in advocacy relations, 122; epistemic trust, 30, 56, 69; in experts, 68, 70, 75–76, 130–31, 150–51; in interactive knowledge production, 149; public trust, 52–53, 119, 128; rationality dependent on, 128

truth, 7, 8

Uganda, 150–53, 158
Uluru ("Ayers Rock"), 13, 58
uncertainty/doubt: epistemology of, 95; as manufactured, 67, 71, 76–77, 116; and sexism, 94; used to legitimize inaction, 60, 178–80. See also climate change skepticism; ignorance
universal justice, 18n28, 50
universal validity of knowledge, 9, 29, 35, 75, 102
universality: in dominant Western epistemology, 52, 82, 142, 148, 206–208; versus (feminist) particularity, 30–32, 40, 142–43, 146; human sameness, 2, 57, 79, 86, 112, 141, 182, 196–97; and male freedom, 31
Uriah Heap, 100, 184, 212–14

"view from nowhere," 17, 29, 65, 107, 110, 201
virtue, 200–17; Aristotle's concept of, 201–202, 210, 212; courage, 203; intelligent knowers, 202–203; prudence, 202–203; requirements/practices of, 203–205; social, communal dimensions of, 203–206, 212; virtue epistemology, 201–202. See also epistemic responsibility; humility
visual evidence, 49
Voltaire, 15, 174
vulnerability, 29, 56, 89, 114–15, 148

Waitt, Robert, 13, 58
"we": analyses/interrogation of, 2–5, 16; entitlement of, 10, 57, 59; as knowers, 8, 153, 161, 174, 190; ontological questions about, 55, 134; responsibilities of, 3, 144; unreflexive use of, 21–25, 33, 48. See also "who do we think we are"
Weigman, Robyn, 31
Western philosophy. See Anglo-American mainstream epistemology/philosophy
What Can She Know? (Lorraine Code), 212
whistle-blowers, 52
"White Ignorance and Colonial Oppression" (Shannon Sullivan), 157
whiteness/white ignorance, 12–13, 57–58, 88, 157, 199, 208. See also affluent white Northern-Western world
Whitford, Margaret, 80, 95
"who do we think we are": "are we the scum of the earth?," 22–24; assumptions about, 5, 16, 55, 152, 179; and epistemic responsibility, 25, 56, 119–20, 173; genealogical analyses of, 3–4; as outraged challenge, 11, 12–13, 56; as professional philosophers, 208–209; radical re-thinking and re-enacting, 14, 28, 59; and social harms, 2–3, 22
"whose certainty/uncertainty?," 181
"whose knowledge are we talking about?," 146, 212
"whose moral beliefs are at stake?," 190
"whose science?," 172, 173
"whose virtue? whose wisdom?," 206–207
"why should a knower care?," 216
"wild children," 42

Wittgenstein, Ludwig, 8, 42, 152, 184
women, 79, 83, 150–52, 198, 208. *See also* feminist epistemology
women as subjects: categorization/stereotyping, 188; concrete specificities of women's lives, 32, 142, 145; diverse experiences of, 198, 208; enabling male achievement, 79; instrumentalization of, 83; marginalization, 63, 80, 111, 163, 186; Ugandan ("Africam") women, 151–53. *See also* feminist epistemology
Woolf, Virginia, 63–64, 83
"world travelling" (Lugone), 85

xenophobia, 216

Yancy, George, 12, 57, 141
Young, Iris Marion, 133

www.ingramcontent.com/pod-product-compliance
Ingram Content Group UK Ltd.
Pitfield, Milton Keynes, MK11 3LW, UK
UKHW041917140426
5217IPUK00013B/199